Crafting a Symphony in Wood

Crafting a Symphony in Wood

The Story of Violin Maker Anton Sie

ELIZABETH ELLEN OSTRING

RESOURCE *Publications* · Eugene, Oregon

CRAFTING A SYMPHONY IN WOOD
The Story of Violin Maker Anton Sie

Copyright © 2016 Elizabeth Ellen Ostring. All rights reserved. Except for brief quotations in critical publications or reviews, no part of this book may be reproduced in any manner without prior written permission from the publisher. Write: Permissions, Wipf and Stock Publishers, 199 W. 8th Ave., Suite 3, Eugene, OR 97401.

Resource Publications
An Imprint of Wipf and Stock Publishers
199 W. 8th Ave., Suite 3
Eugene, OR 97401

www.wipfandstock.com

PAPERBACK ISBN: 978-1-5326-0341-9
HARDCOVER ISBN: 978-1-5326-0343-3
EBOOK ISBN: 978-1-5326-0342-6

Manufactured in the U.S.A.   OCTOBER 13, 2016

This book is dedicated to my children
Sven and Genevieve,
who learned violin and cello from Anton Sie

Contents

Illustrations | viii
Acknowledgments | xi

Prelude	In Gratitude | xiii
First Movement	Indonesian Gamelan | 1
Second Movement	Chinese Pipa | 46
Third Movement	Hong Kong Blues | 104
Fourth Movement	Violin Expert | 160
Postlude	Guitar, in Appreciation | 205
Appendix A	*How to Make a Good Violin* | 213
Appendix B	*Carleen Hutchins' Correspondence* | 217
Appendix C	*Letter from Teacher Anton Piontek* | 222

Illustrations

PRELUDE: IN GRATITUDE

Anton with *mezzoviolin* and African Grey parrot | xvi
Sven Östring with *mezzoviolin* | xvi

FIRST MOVEMENT: INDONESIAN GAMELAN

KiemGiok 1949 | 40
The Menara Kudus Mosque | 40
The Kudus *Klentang* | 41
The Sie Family Home | 41
Dutch Primary School 1949 | 42
One of Four Clowns | 42
Card Magician (signed photo) | 43
Winning the School Ping Pong Cup | 43
School Guitar Accompanist | 44
Teacher Anton Piontek | 44
The Sie Family, 1955 | 45

SECOND MOVEMENT: CHINESE PIPA

On the Ship *Tjwangi* | 98
Physics Class 1956 | 98
Still the Guitar Accompanist | 99

Detail of Anton's notebook | 99

Anton's Degree | 100

Newly Wed Anton and YuXiang | 101

Indonesian Delegation | 101

Anton's Hand Tinted Photo of YuXiang | 102

Anton the China Artist | 102

Anton with his two Daughters, 1976 | 103

THIRD MOVEMENT: HONG KONG BLUES

Anton's *Dancers* Sketch | 155

Anton's *Stradivarius* | 155

Anton's *Flamenco Guitarist* | 156

Anton's *Maestro in the Sky* | 156

Anton's *The Quartet* | 157

Anton's *Paganini* | 157

Anton's *Anne Sophie-Mutter* | 158

Prize-winning NeeZi in Concert | 158

Anton with Jan van den Berg | 159

Anton with Benjamin Hudson | 159

FOURTH MOVEMENT: INTERNATIONAL APPRECIATION

Anton with Carl Pini | 197

The Don Pasquale Opera | 197

On the China Lecture Tour with Carleen Hutchins | 198

Carleen Hutchins Testing one of Anton's violins | 198

The Student Party, 1986 | 199

The Dragon Violin Exhibit | 200

Hutchins' Letter of Recognition | 201

The *Pentaline* Violin | 202

Anton and YuXiang at a Party, 1986 | 203

American Appreciation, 1996 | 204

POSTLUDE: THE GUITAR, IN APPRECIATION

Anton's Logo | 210

YuXiang's Bears | 210

Golden Wedding Photo, 2012 | 211

Relaxing at Christmas | 211

Example of Anton's Guitar Music | 212

Acknowledgments

THIS BOOK HAS BEEN incubating for twenty-five years. I am very appreciative of the widely international variety of people who helped make Anton Sie's dreams a reality.

I would like to express my sincere gratitude to the Sie family for generously sharing details of their lives, and making available many photographs. Anton's daughters offered valuable advice and insights, wisely insisting the story be told in a "lively" way. In the process of writing this narrative the family has moved from being people I admired, to real friends.

An acknowledgement about transliterations of Chinese names is in order. In Indonesian settings transliterations reflect the Fukienese (or Hokkien) Chinese used there, but in China situations the official Mandarin Chinese used in the People's Republic of China. Chinese surnames come first, and personal names, which follow, usually have two syllables, which are here run together to make one word, although an upper case letter is used to show that the name consists of two parts (unless the name is well known, as for example, Mao Zedong). Place names in Hong Kong, however, are the officially recognized transliterations of the Cantonese Chinese of the the once British colony.

I am grateful to the staff of Wipf and Stock for accepting my manuscript, and giving valuable editing advice. Having previously found them a pleasure to work with, it was a delight to learn they had faith in the value of this fascinating story.

Prelude

In Gratitude

"THAT'S AN ODD, SORRY mate, I mean different, violin," Sven Östring's friend observed.

"Maybe!" Sven shrugged. "It's a *mezzoviolin*, handmade by my teacher in Hong Kong."

Sven had just conducted a well-received performance by a small church orchestra in Perth, Western Australia. Although not exactly the Berlin Philharmonic, only a few weeks earlier there had been no orchestra, just a few disparate and mostly mediocre players of various instruments. Now, after hard practice, a lot of fun, even a few disagreements, the once motley group was successfully using the power of music to create joy for their community.

"Hey! Great performance!" Pastor Glenn Townend, Sven's work supervisor enthused, coming up from behind. He clapped Sven affectionately on the shoulder. "Never thought you'd get those people to play that well!"

"Oh, they're a good team," Sven replied, smiling.

"We were talking about Sven's violin," his friend interjected. "It's odd."

"Never thought much about violins," returned Glenn. "Some Italian guy called Stradivarius made good ones hundreds of years ago. But they say his technique's been forgotten, and now no one makes them as good."

"My teacher discovered what makes a Stradivarius so good. That's why he made violins," Sven replied. "Stradivarius also made larger violins like this *mezzoviolin*."

"No kidding? A modern Stradivarius? Well! Well! Well! Someone should tell us about this guy!"

*****

The July heat of Hong Kong in 1981 was stifling. Sven, recently arrived from the cool grey skies of England, was sweaty and bothered. Everyone in the hospital employees' apartment block was away on holiday, at least everyone with children a six-year-old boy could play with. Then Sven remembered a family had just returned. He'd check them out.

As Sven knocked on the door of the ground floor apartment he heard an amazing sound. He knocked harder and the sound stopped. A Filipino boy, clearly several years older than Sven, came to the door.

"Want to play?" asked Sven hopefully.

"Can't," said the boy, staring at Sven in surprise. "Who are you? I have to do my violin practice. You can stay and listen if you like. When I've finished, I can play."

"OK, I'll come and listen," replied Sven, keen not to let his potential playmate disappear.

"If you like!" shrugged Luiji Oliva, shutting the door behind Sven, who ran in and settled himself on the sofa.

"Hey, can you show me how to make that sort of noise?" Sven asked breathlessly when Lui lowered his bow to rest.

"Maybe my teacher could teach you," Lui answered, surprised by Sven's enthusiasm. None of Lui's other friends showed interest in his violin playing.

Sven did not wait for Lui to finish. Without a goodbye, or even a promise of play, he raced two steps at a time to his fourth floor apartment.

Breathlessly bursting through the door he shouted, "Mummy! Mummy! Mummy! There's this boy, and he makes the most wonderful noises! He says I can learn to make them too!"

"Boys don't need to learn to make noises," Sven's mother thought wryly, as she dusted the flour from her bread-making arms. "What sort of noise?" she said.

Sven gave a clear demonstration of violinist.

"I see," she smiled. "We'll talk to Daddy when he comes home." Sven raced back to Lui, eager to share his good news.

"I can! I can! Mummy says I can learn to make that noise too!" he cried, bursting into the Oliva apartment without even knocking.

"Oh! Oh!" said Lui, shrugging with surprise. "Really?" He returned to practicing.

The great day to meet the teacher finally arrived. After a long, hot ride in a crowded bus with Daddy, Lui's father, and Lui, a nervous Sven met the teacher, Anton Sie. But Mr. Sie, a softly-spoken man, was not at all scary. He agreed that if Sven promised to practice very hard every day he would teach him to play the violin. Little did Sven and his family realize the violin teacher's remarkable story and his current desperate need for students. But the man who gave Sven his music and his violin had learned his skills from a Muslim peasant, a Jewish refugee, Russian and Italian composers and teachers, Russian and Chinese Communist scientists, and an expert American violin maker. In music, no man is an island.

*****

Many of the techniques of European violin making were tragically lost during the social, political, and military turmoil of the late eighteenth and early nineteenth centuries. Not until the twentieth century were some of these techniques rediscovered, due to the determined persistence of a few dedicated scientists and musicians.

Anton Sie told the author about Leonardo Fioravanti, a perceptive Italian who wrote in 1573 that an instrument maker needs to be a painter, a smith, a master woodworker, a musician, and an alchemist. Anton Sie embodies these multifaceted skills. But more importantly, his story illustrates the inter-connected, international debt humans owe each other. His story is an inspiration for all who are willing to work towards goals they believe important, to learn all they can, remain focused on the task, and use their knowledge and ability to benefit to the world. Anton Sie's story honors his indomitable spirit and remarkable determination, the truths of science, and the joys of sharing music. Despite a majority belief that the art of the great Cremona luthiers had been lost, Anton did not accept this pessimistic verdict, and instead showed that all that was lost was a willingness to work as carefully and meticulously as the ancient masters.

This story has symphonic form: four movements, with prelude and postlude. Conversations reflect real situations that capture the meaning and intent of events in Anton Sie's life, and, although they do not claim to be historical transcripts, they reflect his ideas, and often own words, for example in the closing scene of the book. For those interested in the science of Anton Sie's work, appendices are included, but a wealth of data about his beautiful violins has reluctantly been omitted in the interest of general readership and space.

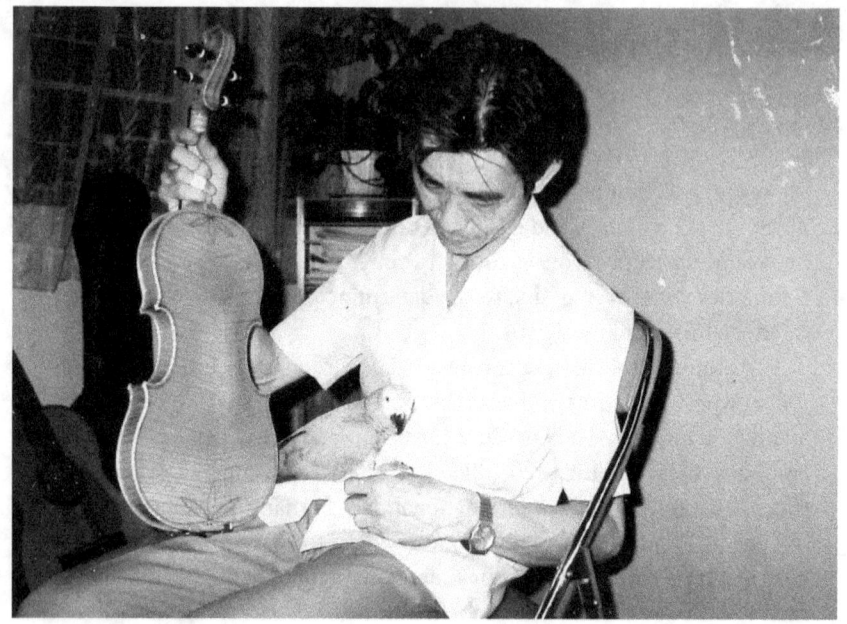

Anton Sie holding the newly completed *mezzoviolin* that Sven Östring now plays, while enjoying the company of his African Grey parrot.

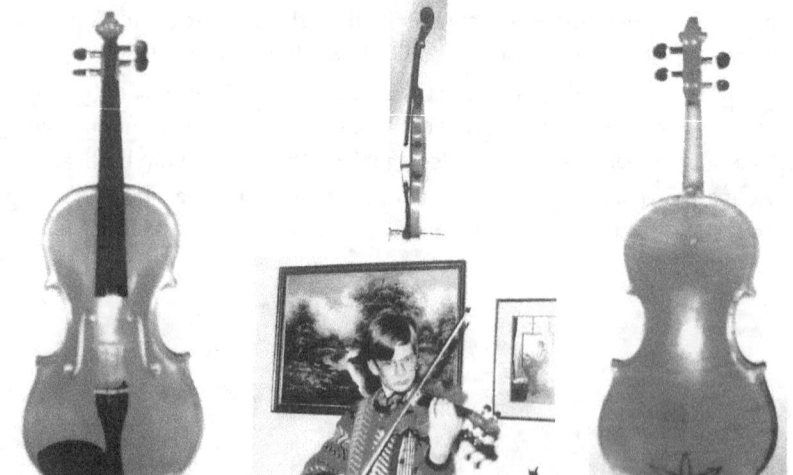

Anton Sie's picture of Sven Östring with his "Babingtona" *mezzoviolin*, 1990.

First Movement

Indonesian Gamelan

KiemGiok skipped happily beside his father as they walked towards a small *klengteng* (Chinese temple) in the provincial city of Kudus, Indonesia. Their route took them past the beautiful Menara Kudus Mosque. He gazed up at its elegant red brick minaret, hoping he would hear the *bedug* drumming its call for local people to come to prayer, but it was silent. The Kudus drum tower was unique; nowhere in the country was there anything like it. KiemGiok thought the calls of the *muezzin* in other mosques were rather sad, but the drum rolls of the Menara Kudus Mosque *bedug* sounded cheerful and important. The shape of the minaret was attractive, so different from the squat domed shape of the mosque beside it.

"Father," he ventured as they walked along, "why is the tower so different from the mosque?"

KiemGiok was very excited about this adventure. He could not remember ever doing anything like this with his father, Sie TjwamKhing. Father was always busy with his work as a bookkeeper for the local cigarette factory. KiemGiok was proud of his father's work. He knew the *kretek* cigarettes made in Kudus were the best in Indonesia, and probably the whole world. *Kretek* cigarettes smelled nice, made of a mixture of tobacco leaves, clove spice, and sometimes nutmeg, cumin, or rubber resin.

Father Sie took time to answer his six-year-old son. So long, in fact, that KiemGiok thought he must have done something wrong, the last thing he wanted on such a special occasion. He was trying to figure out what he should apologize for when Father's voice soothed his anxiety.

"Kudus is an old city, son," Father began. "Our family has been here for six generations. They came from China, a country far away over the ocean. They came because men were fighting and killing and destroying in China. These men wanted what belonged to Chinese people. They were Englishmen trying to force our people to buy opium, terrible stuff. Foreign mud we called it. It makes men crazy, unable to work and think."

Father Sie was silent for a long time, puffing quietly on his *kretek* cigarette. He remembered the horrifying stories of the nineteenth century Opium Wars, stories that passed with terrible urgency through successive Sie generations. He knew that China, convinced of its superiority, had refused to trade with the pestering English barbarians. China was not surprised that the red-faced, red-haired Englishmen wanted their tea, but the only thing China wanted in return was cash, good solid silver cash. The wily English, however, humbled the great Emperor, and found something that even if he did not want to buy, many of his people did. Opium poppies grew freely in India, and their sap was easily shipped to the southern coasts of China. Despite proud imperial edicts against its use, opium proved extremely popular with people burdened with relentless hard work, or merely bored with leisure. They paid cash for the stupefying effects of poppy sap, cash that could then buy tea. Opium sap produced an addiction with terrifying withdrawal effects, and ensured a relentless market for it dubious benefits. Until the Emperor forced the issue, and confiscated a mountain of "foreign mud" chests stored in godowns beside Guangzhou harbor, the English did a roaring trade. The enraged English responded with gunboats. No one in China was too concerned when a tiny, obscure, and virtually uninhabited island, poetically named Heung Gong ("fragrant harbor") at the mouth of the Pearl River in Guangdong province was captured. But the conquerors went pillaging up the coast of China, and when they reached Fujian province the Sie family decided it was time to use the honored Chinese method of avoiding trouble: migrate. Fujian Chinese were boat builders and sea traders who used their skills to travel down the coasts of China, Vietnam, and the Indonesian islands to reach the north coast of Java.

Suddenly Sie TjwamKhing shook himself out of his reverie, took a deep drag on his cigarette, and resumed his story.

"Yes, opium. It's very bad stuff, very bad. Son, you must never touch it, never ever, ever. It destroys your mind. It destroys your will. Our ancestors were real men, and they risked their lives to come to this new country. They were determined to be honest and good citizens. We would be independent and useful to the people who had given us a new home.

"But we learned that our new country was not a free country after all. Like China, it too had been conquered by Europeans. The conquerors of Indonesia were Dutch people, the people who are in charge of our country now. They are strong people. No one dares defy them. Some people are never satisfied with what they have. They always want more; they want what other people have. It's bad, very bad, to be like that. People should be content with what they have.

"The Menara Mosque belongs to Indonesian people. It's very old, much older than the Dutch buildings here, very special, and no one really knows who built it. The mosque has the tomb of one of the nine heroes of Java, called the *Wali Sanga*. In fact, two of the *Wali Sanga* are buried here in Kudus. Our city is very special. There are many stories about how the tower was made: some even say it was miraculously built in one night, before dawn!

"But strange things are happening in our world. Where Dutch people come from, far away in Europe, there are people who are even stronger than them, called Germans. They have marched into the Dutch country and taken control. It seems too strange to be true. People at work are excited, because they think they can now become free from the Dutch, free like they were hundreds of years ago, long before our family came here.

"Ah, but there are even stranger rumors, scary ones. We hear there are other people who want this country. These Japanese are very good soldiers. My friends at work are happy. They think if the Japanese come it will help our country to be free. But I am worried. I remember what the English did in China. They said they wanted tea, but they gave us poison, stole our money, and destroyed our palaces. They even destroyed the Summer Palace of the Emperor! I don't think the Japanese want to help anyone in Indonesia. They only want something for themselves. Why don't the Japanese stay in their own country?"

KiemGiok trudged along silently. He was very confused. He never expected such a long reason for the mosque tower to look different from the mosque. Never before had his father said so much to him, or sounded so serious. Despite the burning hot tropical sun shining down on him, he shivered. He forgot about the Menara Kudus Mosque tower, and simply hoped those j-something people would decide to stay home.

Presently they reached their destination, and father's long speech was forgotten. Compared to the Menara Mosque the little Chinese temple seemed very small, but it was bright and cheerful. Its roof was golden-orange tiles, and six bright red pillars supported it. A pair of green and white porcelain lions guarded the entrance, and a large green and gold dragon surveyed the courtyard. This area during Chinese New Year celebrations was crowded with people enjoying stalls selling various Chinese snacks, as well as bazars hawking toys and trinkets. Although there was no bazar around the temple this day, it was the venue for the *wayang gamelan* shows. If his speech had surprised his son, Father Sie was even more startled by his son's first comment when they arrived at the temple grounds. He was sure either the puppet show, or the rowdy bazar was what KiemGiok wanted to see.

"Oh! Oh! Look Father! Look at those music things! Aren't they beautiful!" KiemGiok exclaimed. "Oh, thank you, thank you Father!" the small boy jumped around with excitement. Then he giggled. "They look like cooking pots, don't they? But I love those music things! I'm so glad you let me come here!"

The previous day, when KiemGiok brought home his first school report, he was amazed by his parents' obvious happiness. Of course, as well-mannered Chinese they would never tell their small son they were astonished at how good his report was, but neither could they conceal how very, very pleased and immensely proud of him they were.

"You will bring honor to our family," Father had intoned seriously. "Keep working hard."

"He even got one hundred per cent for Dutch language!" his mother whispered in unmistakable pride.

But KiemGiok still couldn't believe what happened next.

"Father," he said, stunned at his own audacity, "Father, have I . . . have I been good enough for you to take me to the *gamelan* in the park tomorrow?"

There was utter silence, finally broken by Father's disapproving voice. "How do you know there's a *gamelan* show? That's such frivolous stuff, son!"

"I . . . I don't mean the *wayang*, the puppets, father," KiemGiok stammered. "I . . . I . . . I mean the sounds. I want to see how they make those sounds. My school friends told me about it."

"It won't hurt," his mother, Tjan SingNio said, suddenly and unexpectedly. "We should encourage him." She gestured towards the report lying on the table. "You don't have work tomorrow, and I'll get KiemLiang to help me with the chickens."

Father shifted awkwardly in his chair. "It's not usual," he said. "The boy shouldn't need inducements."

"But it would be good for both of you. I know you've been worried about the rumors of war flying around here. Yes, it would be good for you, too."

Father sighed. "If you can get all your work done before we eat tomorrow morning, we can go," he demanded of the wall, rather than the small boy staring wildly at his face. His wife smiled triumphantly. Of course, KiemGiok was out of bed before dawn, before the most reliable rooster could start his crowing. Chores were completed well before the first meal.

"Can we get close?" KiemGiok asked his father, tugging at his sleeve. "I really want to see how they make the sounds."

Father agreed, and presently he enjoyed explaining the various instruments to his son. He too loved music, and although he did not want to show his pride, he was delighted that KiemGiok obviously shared his interest.

"See all those things that you say look like cooking pots? They're *bonang*. See that drum, it's called a *kendhang*, and it's played with both hands. Those things with metal strips over the bamboo pipes are *gendér* and *slentham*, and the metal strips over a carved box are what we call a *peking* or *saron*. All of them are hit with mallets, like the *bonang*. Oh, and look at that beautiful *kenong* suspended in its gold embossed frame!"

KiemGiok looked at his father, eyes bright with joy, and a sudden realization. "Father!" he exclaimed. "You like these noise things too! You like sounds too!" Then he giggled. "But father, they really do look like cooking pots, don't they!"

Father Sie grinned sheepishly. "Yes, son, I do like music. And yes, they do look like cooking pots. But you know, there are musical instruments that are even more mysterious, that make even better sounds than these."

Maybe there were, but as the musicians, clad in their colorfully patterned *batik* robes, began to play, KiemGiok was lost in wonder. Sounds were such beautiful, such truly magical things. What makes sound, good sounds and bad sounds, he wondered? Each one of the *bonang* made its own sound, a sound that could be repeated every time it was struck by the musician's mallet. Did the sound come from the wooden mallet that hit it, or the *bonang*, or even the stand the *bonang* was resting on? Sounds could be mixed up to make a happy feeling, or a sad one. There were so many things to find out about sounds.

When father and son walked home late that afternoon, just as they reached the Menara Mosque the *bedug* began insistently calling the faithful to prayer. Although earlier he had so much wanted to hear that drum, the unexpectedly loud sound of its call close by seemed strangely ominous. KiemGiok shivered, and was glad his father walked beside him.

*****

Months rolled by. The Japanese that had worried Father Sie landed and over-ran Indonesia. Like the Dutch before them, they made the area around Semarang and nearby Kudus the center of their activity. The Indonesian government made noises about things getting better, but no one in the Sie household saw any improvement. There was merely a lot less food to eat.

One morning, KiemGiok, now in his third year at school, jumped up and down impatiently, waiting for his mother to finish cooking breakfast. "Stand still!" she commanded irritably. "You make me nervous! Drink your tea if you are in such a hurry." Mother was always cross these days, so he paid little attention to her bad temper.

"Where's father?" he asked, as he sipped the small cup of hot water handed him. Mother never put any tea leaves into his cup these days. "Why doesn't he have breakfast with us any more?"

"He's busy," mother snapped. How could she tell her son that her husband now left each day while it was still dark to avoid being seen on the streets? He tried to find something to eat along the way to avoid sharing the meagre rations of the family.

Mother scooped half a fried banana from her pan and dropped it into KiemGiok's bowl. "Only half?" he whined as he fanned the hot fruit ready to put into his mouth. "Only half again? Why don't we ever have *nasi goreng* now?"

"Be quiet!" Mother snapped. "You're lucky you have that! The other half is for your brother, and you know it!" She gave him a quick slap.

Yes, he was lucky, very lucky, and KiemGiok did know it. Half a banana, even when cooked in lots of coconut fat, did not stop the tummy rumblings of a hungry boy, but many of his classmates did not get even that before they left for school. His family lived on the edge of the city, close to villages and farmland, where they could grow things to eat, and knew friends who grew food. But his schoolmates who lived in the city were not so lucky. They begged him to bring food from his mother's small shop. But how could he steal from his own mother? Stealing was an unthinkable crime in the Kudus community. The banana tree at the side of their house had a huge bunch of bananas on it, but he knew most of them went to mother's shop to earn money for other food for their family. Yet somehow, mother always had half a banana every morning for her two boys.

School was strange now. Old Herr the Dutch language teacher had disappeared, no one seemed to know exactly where. They heard of "those" places where once powerful white people now lived in terrible conditions, behind barbed wire fences. KiemGiok was sad because he liked the old man, and now there were no more Dutch lessons.

Until the soldiers arrived, children always chased each other and played games as they walked to and from school. Now they learned to avoid the attention of Japanese soldiers at all costs. If a child was in the path of these marching men with their bayonetted guns, he or she could be beaten mercilessly while the Japanese yelled incomprehensible words. Sometimes the soldiers used a few Indonesian words, but they were always bad words, insulting words, words that never helped children know what they had done wrong. Once, to the horror of KiemGiok and his friends, their friend SiouYin[1] chased a chicken into the street just as a group of four

---

1. A pseudonym.

soldiers stomped by. The soldiers grabbed him and tried to take the chicken. When SiouYin clutched it tenaciously he was beaten brutally, kicked many times, and finally, when he did not get up when commanded to, one soldier cut off his head with his bayonet, and kicked it down the street. Terrified, KiemGiok and his friends slunk through the shadows to school, too frightened to tell their teachers what they had seen. They decided they must have done something very, very wicked for such a dreadful thing to happen to their friend.

KiemGiok had plenty to keep him occupied after school now, and no time to play. His job was to feed the family chickens (those they still had). This meant looking for chicken food and finding any eggs they laid. Finding eggs meant mother had something to cook for the evening meal, so KiemGiok looked very, very hard for eggs. Yet almost every night another hen disappeared. Mother complained bitterly about this, but Father would say, "Be patient, woman, the people are starving. I'll look for some chicks." But there were either no chicks, or he never looked.

Father was concerned about KiemGiok's looking for hen food. "It keeps him out of mischief," he said to his wife, "but I don't like him out on those roads. I wish you would be more careful. He's almost eight now, and I've heard the Japanese use very young boys to train for their army. They take them away to their own country, or somewhere."

"Surely not so young!" his wife replied, frightened and shocked. "What next will they do!"

"Take you or your daughter to meet their desires, or your shop!" retorted her husband. "You know as well as I do what we've heard."

KiemGiok was listening hard. He did not understand all his parents said, but he understood enough to determine that he would be very careful, very, very careful. And he would look even harder for hen food. He was not a woman, so he was not worried about helping soldier desire, although it did not seem a good thing to do. He remembered the day he went on a message for his mother, and saw a young woman with two soldiers. She was lying on the road weeping and whimpering, "Stop, stop!" but the soldiers just lay on top of her and laughed. Perhaps, KiemGiok thought, he could make a little house for the hens, so they would be safe at night, and no one could steal them. He could pretend he was just playing, and maybe soldiers would not notice what a boy was doing.

KiemGiok pinched himself to stop his scary thoughts. Half a banana does not take long to eat, and soon he was on his way to school. Suddenly he heard angry shouts and screams. He quickly crossed the road into the shadows, and crept along slowly. A large crowd gathered in front of the house of

Liem SiongSioe[2]. Japanese soldiers were shouting at Pak[3] Liem. One pinned him against the brick fence. Three hammered at the hinges of its beautiful wrought iron gates.

"How dare you take my gates!" Pak Liem screamed. "What use are they to you? You steal our oil to run your stinking trucks! You take our oil to your own country. But what use are my gates to you!"

A Japanese soldier yelled, and hit Pak Liem's head with the butt of his rifle. Pak Liem crumpled to the ground, moaning, as blood poured from a large gash across his cheek.

"Don't be crazy!" a voice in the crowd urged. "They take anything, you know that! If you resist they'll kill you!"

"My beautiful gates, my beautiful gates," moaned Pak Liem, pressing the sleeve of his white shirt against the gory gash on his face.

"They took the iron gates from the Menara Mosque yesterday!" another voice yelled. "They respect nothing!"

With a final hurl of their mallets, the Japanese heaved the gates from their hinges, and they clattered on the dusty road. The soldiers yelled at the crowd and pointed animatedly at two young men. No one moved. In swift strides the soldiers grabbed the men, and roughly shoved them towards the fallen gates. Bewildered, the men did nothing. The soldiers picked up the heavy gates and threw them at the men. The decorative scrolls of one gate caught one man on the arm, tearing a gaping hole. The other crumpled beneath the gate, his head bent at a strange angle, and did not move. A soldier strode across and kicked him. There was no response. Cursing, the soldier grabbed two more men and gesticulated unmistakably that they were to carry the gates. As the luckless porters trudged down the street dragging their heavy burdens, the crowd rapidly dispersed. All that remained of the altercation was a very still form on the road, and Pak Liem leaning against his brick wall moaning "my beautiful gates, my beautiful gates," as blood from his wound steadily dripped down on to his bloodied white shirt.

Sick with fear, KiemGiok crept along the shadows to school. But he could not concentrate on his schoolwork, or stop remembering his friend SiouYin. He decided that before he went to find chicken food he should tell his mother what had happened. He knew instinctively he must do things differently, and perhaps she could tell him how. Her little shop selling basic food was not far off the route home, anyway.

---

2. A pseudonym.

3. *Pak* literally means "uncle", but is a polite Indonesian term of address for older men.

But as he came close to the shop he was horrified to see his mother was no longer there. Standing at the counter presiding over a meagre array of bananas, a few tins of food, and a precious barrel of rice, were two strange, dark skinned men, their heads shaved bald. They were wearing Indonesian clothes, but surprisingly did not have *songkok* (Indonesian caps) on their bare heads. KiemGiok was paralyzed by fear. Did Japanese soldiers take mothers as well as gates?

While he was considering whether to run and tell his father the terrible news, he heard the men talking in low voices. Their voices were soft, not the voices of men. He listened harder, and suddenly he knew those voices.

"Mother!" he exclaimed, in a hoarse whisper. "Mother! Why are you looking like that! What happened to your hair? What happened to your face? It's black and dirty! Why are you wearing those awful clothes? They look like Pak Budiono's[4] dirty old ones!"

"Shh!" mother motioned with her finger against her lips. "I will tell you later. Go home at once! Look after the hens."

"Yes, Mother." He was so relieved to find he still had a mother he forgot about Pak Liem and the gates and ran home as fast as he could. The front door was swinging wide and hens were pecking around on the tiles inside, where they most certainly should not be. Mother must have left home this morning in an awful hurry, he thought. He shooed the hens out and, with water from the well in the backyard cleaned up their mess. That will please Mother, he decided. Then he began his routine task of finding hen food.

Why would mother want to look so horrible? She was a beautiful woman, he knew. And his sister, why did she have such a dirty face and look so strange? KiemLiang, his sister, was forced to leave school as soon as she could read and write, despite being a good student. She often looked sad because he knew she did not enjoy working in their mother's little shop. He was so busy pondering all these imponderables that he almost missed a nest with three beautiful eggs. Three eggs! A tremendous find. They could feast tonight!

Mother was bending over the grass-fired stove cooking the precious eggs when father arrived home. As she straightened up and he saw her shaved head and dirty black face he gasped.

"It's me, truly, me, TjwanKhing," she said softly, with a giggle. "Don't stand there looking so horrified. You told me this morning to be more careful!"

"But your beautiful hair! Why?"

---

4. Again, a pseudonym

"You know very well why! Japanese soldiers were marching up our street after you left for work. There was trouble at the Liem house. One knocked loudly on our door. We hid, of course, in that big pile of banana leaves you were smart enough to make at the back. He kept shouting "any women", or something that sounded like that. He was poking around in the house. As soon as he left we ran to the shop. We must stop looking like women the soldiers might want, I decided. Are we ugly enough?"

"You are very ugly," her husband said, shaking his head. "No one would ever want you! But where did you get those clothes? I've never worn Indonesian clothes, and never anything so dirty!"

"I keep old clothes at the shop for cleaning, and people give me old clothes to sell as rags," Mother explained. "I just took what was in the rag bag. The older the better I thought!"

Her ruse worked. She and KiemLiang were never molested. They survived. Many did not.

*****

In 1945, KiemGiok's tenth year, conditions became extremely difficult for the Kudus community. Japanese soldiers were more and more brutal towards the Indonesian population. A few wealthy people in Kudus had small radios secreted in their houses, and they knew that in Europe and the Pacific region Americans and "allies" (whatever they were) were wining battles. Perhaps the soldiers sensed the surge of optimism that accompanied these radio reports. They beat mercilessly anyone suspected of having a radio, and anyone with food. KiemGiok's mother was in many dangerous situations that caused her husband great anxiety.

But the Chinese population felt particularly vulnerable. With more and more Indonesian people talking about freedom from foreigners they felt they could trust no one. Night after night, late into the night, KiemGiok heard his father talking with friends in very low voices.

Suddenly one night it was all action. Mother arrived home with an old basket containing as much of her stock from the tiny shop as she and KiemLiang could carry without attracting attention. KiemGiok caught the remaining two hens and tied them together. He put dirty trash over banana leaves covering a precious bucket-load of rice. Father rolled up their bedding, and handed around black clothes for everyone to wear. Mother set the table as though they were about to eat, and made makeshift beds from the remains of the banana leaves in the backyard.

"So the Japanese won't know we have run," she answered KiemGiok's wondering eyes.

When the sun set and darkness fell like a black velvet curtain the family trudged silently through the moonless tropical night. One by one, two by two, other dark silhouettes joined them, but no one spoke. Despite the blackness of the night KiemGiok recognized some of his father's midnight friends.

Their destination was a small, deeply secluded Chinese village. One village resident, who had worked for Caltex oil before Japanese seized its oil fields and refineries, stashed several drums of petrol in bushes around the village. With the support of the villagers and their guests, he agreed that on cue he would pour a ring of precious petrol around the village; if soldiers threatened they would set fire to this ring, and die an honorably chosen death by their own hands, rather than be captured and tortured by the hated Japanese.

Days turned into weeks. Food disappeared from the village, despite nightly forays by the men. Not a rat, not a snake, disturbed their sleep: all had long since been eaten. With a permanently rumbling stomach KiemGiok longed for just one bite of banana. He was proud his mother proved best at finding wild fruits in the surrounding forest. There were no hens to feed as all were eaten. Sometimes he and other children found wild birds' eggs in the fields, but that often caused fights over who got the eggs.

One evening, about six weeks after arrival, the weary, hungry people were sitting outside their makeshift shanties, hoping for a breeze to ease their lives. Idly, one man picked up a cooking pot and hit it. The metallic sound rang through the night.

"Quiet!" commanded the village leader. "Do you want to die?"

"But we can make music," said KiemGiok, shocked by his own boldness. "We can get all the pots and make *gamelan*."

"Why not!" a voice in the darkness responded. "Why not? The boy's right! We might as well make music before we die!"

Some giggled uneasily, but several men got up. They emerged from shacks carrying a motley collection of cooking pots. KiemGiok scrabbled around in the darkness and retrieved an empty petrol tin he had found. Tentatively he began rhythmically strumming. A wok-turned-gong rang triumphantly through the night. Suddenly a strangely melodious cacophony of banging pots and pans, with a petrol tin *kendhang*, shattered the stillness of the night. Women came out of the shacks laughing. One even danced; most had no energy for such frivolity. For a couple of hours this Chinese village in Japanese-ruled Indonesia abandoned restraint.

Next morning the villagers were horrified to see unknown Indonesians wearing colorful sarongs and classic black *songkok* approaching their hovels.

"You were making a lot of noise last night," said one gruffly, frowning severely.

"You look hungry," his companion added, kindly.

The Chinese were silent. Why had they been so foolish to attract attention with their noise? Who were these men? Silently, with hollow eyes, they stared at their visitors.

"Were you celebrating?" the Indonesians demanded.

Silence. A hacking cough convulsed an elderly man.

"Well, if you were celebrating, you're a bit late. The Japanese left a week ago. It's Indonesia for the Indonesians now! Hurrah! Hurrah! Hurrah!"

The Chinese villagers were stunned. What were these unsmiling men going to do to them now? Were they to be rounded up and sent to some terrible, nameless place like the White Dutch?

"We have some rice the Japanese left. Would you like some?" the kindly man suddenly offered with a broad smile.

Would they! The eager faces were eloquent. "It's Indonesia for the Indonesians, even Chinese Indonesians," the man grinned.

"You have food, really?" asked the villagers.

"Come with us."

After a slender meal of rice and a few vegetables with their new friends, the Chinese Kudus city dwellers bundled up their tattered possessions and headed home. They were no longer willing to be a burden on the Chinese village. Their kind Indonesian friends allowed them to borrow an old bullock cart to carry those too weak to walk. "Bring the cart back when you are all well!" they said. "May Allah be kind to you."

Ah! Indonesia for the Indonesians really did mean Indonesia for Chinese too.

*****

KiemGiok walked, even skipped, beside the bullock cart to Kudus. He struggled to carry two new acquisitions. The first was the realization that although he was a citizen of Indonesia, and most of his and his parents' friends were Indonesian, he actually was not Indonesian. Some of the people in the little Chinese village had spoken a strange language, not the *bahasa Indonesia* that KiemGiok and his family, like most people in Kudus, spoke. When he asked about this, his mother told him the villagers were speaking a dialect of Chinese known as Fukienese. Even though his father told him their family came from China, he thought China was part of Indonesia. He was happy to be Chinese, but he had a faint sense of loss when he discovered he was not quite the same as his friends, not quite Indonesian.

His second acquisition was a gift from Pak Ong.[5] Pak Ong was the oldest man in the Chinese village, but he developed a close friendship with KiemGiok. Their bond was music. The old man possessed an ancient two-stringed *erhu*. Once he had been a skilful performer on this traditional Chinese instrument, but now his arthritic fingers refused to place themselves in the right positions. Yet when he showed KiemGiok the rudiments of playing, he was astonished at how quickly the boy mastered the basics. He loved his *erhu*, but sensed KiemGiok would love it even more.

They soon found Kudus had not returned to normal, even though the hated Japanese had left. The Dutch returned to reclaim their lands, but to their immense surprise and hurt they discovered they were not welcome. Sukarno, a rousing orator, convinced his compatriots that Indonesia was indeed meant for Indonesians, not for Dutch. Despite a general agreement about independence, various Indonesian factions with various political viewpoints clashed with each other, and the Dutch fought all of them tenaciously to regain their former colony. The country was plunged into several more years of civil war.

The Sie's small Indonesian-style house was unharmed by the Japanese. It merely needed a thoroughly good clean before they could settle back into it. Father Sie returned to the *kretek* factory. His wife was immediately busy: everyone was hungry so any food she found for her small shop was quickly snatched up. Her business flourished. The Sie family became adept at avoiding trouble with the warring factions.

KiemGiok returned to primary school, and studied hard. He loved to fly large kites made from old newspapers, bamboo sticks, and string wheedled from his mother. She kept leftover rice porridge for him, which, when congealed, made excellent glue to construct lumbering but effective kites. He and his friend Liem KianGie[6] became experts at flying their paper creations. Their homemade kites flew like birds, but they enjoyed the brightly colored ones sold in the bazar beside the *klengteng* at Chinese New Year. A challenging "game" was to see who could wheedle the most money from parents' pockets and get the biggest kite.

Marbles were another favorite game. KiemGiok and his friends were experts, and KiemGiok the undisputed star of marble contests (the reward was a large collection of beautiful marbles). But this was a dubious skill: loss

---

5. a pseudonym
6. a pseudonym

of a treasured marble often made other boys angry. Words were not their only weapons. Father and Mother Sie strictly forbade fighting, so KiemGiok developed creative explanations for large black bruises, bleeding cuts, and grazes.

In spare moments KiemGiok played his *erhu*, trying to unlock the secrets of its sound. His good scholarship and interest in the *erhu* attracted the attention of his headmaster, Chen XiJiong. Chen taught him the basics of *erhu* playing. KiemGiok gathered old cooking pots, and with his friends developed a *gamelan* group. They had great fun making noisy but creditable *gamelan* music. Headmaster Chen made the most of these games, and encouraged the boys to develop a comic performance the school hosted for public concerts.

Finding food for the family hens remained KiemGiok's responsibility. Scraps were given the fowls, but they needed more if they were to lay well. KiemGiok kept a sharp look out for grain a careless farmer, or merchant, might spill on the road, though such lucky finds were rare. Snooping around village pathways looking for green plants and grass seed heads gave him wonderful opportunities for exploration. One day he met a slightly older Indonesian boy also gathering chicken food. At first they were angry competitors and wary of each other, but they started talking, and eventually became firm friends, pooling and sharing their finds. Budi Halim[7] introduced KiemGiok to the games Indonesians played in their villages, especially riding water buffalo in the rice fields. Indonesian peasant farmers worked hard, but made life as pleasant as possible. They teamed up to get rice fields ploughed and planted in the age-old, back-breaking way, bent over from the waist, one tiny rice shoot at a time. When the buffalo were not in use the farmers were happy for children to play with the patient and placid beasts. Crime was unheard of, and children played happily and safely.

It was out clinging to the back of the Halim family buffalo with Budi that instigated KiemGiok's lifelong interest in birds. But it almost brought the boys' friendship to a speedy end.

One day, as they plodded sedately along a narrow path between paddy fields, a flock of pink birds rose twittering into the air. Suddenly Budi jumped off the buffalo's back, grabbed the stick used to goad the patient beast, and began swinging it round in a frenzy. To KiemGiok's astonishment huge clouds of beautiful birds rose in the air, and Budi was like one possessed as he hit at them.

"Got ten of them!" he shouted triumphantly as he ran past his friend.

---

7. a pseudonym

"How could you!" screamed KiemGiok. "They're so beautiful. Why kill them? You're a beast!"

Budi stopped short. "Why wouldn't I? Don't you know what they are, you idiot?"

"Of course! They're sparrows. Beautiful Java sparrows."

"No they're not! They're horrible pests! A flock would eat all my father's rice in a day. Father said he'd got rid of them all from here, but this lot must have escaped. Look! Some are coming back!" he raced off swinging his stick wildly to dislodge the birds, shouting, "Tell Father!"

KiemGiok was horrified by this wanton avicide. Goading the buffalo with his knees he hurried the gentle beast along the trail. By the time Budi tired of chasing Java sparrows from the paddy field his buffalo was a kilometer away. His gesticulations to return his transport were ignored, and KiemGiok watched gleefully as Budi trudged through the sticky paddy fields to catch up with him. No one can run through black mud.

Budi's eyes were blazing when he reached KiemGiok. "Get off my buffalo!" he shouted. "You . . .! You . . .! You thief!" His hands clenched in tight fists around his brandished stick. "Get off!"

Budi was big and strong. KiemGiok did not argue. The boys stood on the path glaring at each other, fists raised, the buffalo snorting between them. Suddenly KiemGiok turned and walked away. Budi dropped his fists.

"Look," KiemGiok said quietly. "There's a pretty bird over there. What is it?"

Budi took a long deep breath. "That one's OK," he said. "It's a pond heron, and it's in breeding plumage. Other times they're just dull grey."

Budi kicked a rock in the path. "And I suppose you're OK too. You're just dumb, that's all."

KiemGiok grinned. "I know sparrows are supposed to be pests. But I love birds."

"Shall I show you where I found a blue kingfisher?" asked Budi, "They're really beautiful." KiemGiok nodded.

They climbed back on the buffalo and plodded to a small sandy bay beside the nearby stream. After a few minutes' patient waiting a piece of sapphire blue dropped from overhanging branches, and sat on a rock. Suddenly it darted into the stream, emerged with a silver fish in its beak, and throwing its beak high, swallowed it head first.

"Neat, eh?" grinned Budi proudly.

"Beautiful. My Grandfather has a mynah in a cage, and it talks. Grandmother looks after my brother, and the mynah copies him," said KiemGiok.

"Yeah, they talk real well. My father says Indonesia has more native birds than any other country in the whole world!"

"Wow! Really? That's pretty neat! I love birds. They make such interesting sounds."

Budi laughed. "There you go again! You're always going on about sounds and noises! Do you really think they're important?"

*****

There was one bird about whose function the boys were totally agreed: the domestic cock. When their friendship reached the completely trusting stage Budi shared the family secret: several members owned gamecocks and they were all fanatical followers of cock fighting. He invited KiemGiok to watch the fun. Soon KiemGiok was also an ardent follower. The Indonesian government, however, banned cockfighting, and meted out severe penalties, but in rural villages it was easy to find secluded places and indulge this favorite pastime. There was no gambling, merely the excitement of winning. KiemGiok never owned a gamecock, but he found every opportunity to join Budi when there was a fight. Games were well attended by scores of excited people, often near a market where they could melt into the stalls if there was a police raid. Because both his parents worked long hours it was many months before they discovered how KiemGiok's chicken-food-gathering trips often ended.

One day when a game was at its most exciting, there came an urgent double "Snakes! Snakes!" Grabbing their valuable birds, the owners melted into the market alleyways, and headed for the poultry bazaar. Budi and KiemGiok ran for their lives, knowing full well that being underage would not save them from severe punishment. When the police arrived a few minutes later all they found were a few men playing bowls with coconuts. Despite their innocent appearance the police took them for questioning, and for good measure a few nearby market stallholders. But their efforts were in vain; all the men were eventually released. Greater precautions were now taken when games were held, but despite all care, police raids continued.

"Snakes! Snakes! Snakes!" called the lookout with triple urgency late one afternoon. A stampede followed, but an unlucky cock owner ran straight into the police. No point trying to pretend that a beautiful rooster with knifelike spurs tied to its legs was about to become dinner for the family. With such compelling evidence the police began a search of the area and arrested for questioning anyone on the road to Kudus.

"Come home with me," Budi urged.

KiemGiok dared not refuse. The tropic night would soon fall, and trying to skirt around the main road through the forest would take hours. The risk of getting lost, or meeting snakes and other dangerous creatures meant it was not safe after dark. Budi lived in a village harmlessly far from the main road, and it proved rather fun staying with an Indonesian family. KiemGiok enjoyed eating *mutarbark* (halal meat wrapped in paper thin bread) with them.

Next morning Budi's father offered the family buffalo to get KiemGiok home in time for school. Pak Halim took him to the edge of Kudus, and let him walk the easy distance home. KiemGiok entertained wild ideas of pretending he got lost looking for chicken food, but as soon as he appeared in the doorway he was stopped short by his mother's blazing eyes. Wordlessly she picked up a *sapu lidi* she had made (a bundle of coconut leaves tied like a broom). Too late, KiemGiok realized her intent, and although he made to escape, she seized him and beat him severely around his bare legs. Only when she had satisfied her anger and let him go did she speak.

"You idle disgrace to our family! Yes! Yes! I know what you have been doing! Father went out looking for you, and heard on the public radio about the police raid. If you ever go near a cockfight again, you will get worse punishment!"

The pain was too severe for KiemGiok to risk trying to defend himself (lest he dissolve into unmanly weeping). Now he had both police and mother to contend with if he wanted to watch cockfighting. Perhaps it was not worth it after all. Well, until next time! At least mother had not said anything about not seeing Budi again. Later that afternoon the boys met and searched most diligently for hen food. Budi offered to take KiemGiok for a buffalo ride.

"The villagers have decided to stop fights for a while till the police get interested in something else," Budi told KiemGiok.

The ride was meant as a treat, but KiemGiok's legs were so painful he had to curtail the fun. Budi didn't ask; the welts on his friend's legs were obvious and he guessed what had happened. But when cockfighting resumed a few months later, Budi and KiemGiok could not resist joining the spectators, although now they watched fights only on mornings they were free from school, with no risk that KiemGiok would not be home for his evening meal.

*****

When KiemGiok was thirteen and in the final class of primary school Headmaster Chen gave him a sealed envelope for his father.

"What's this all about?" Father demanded when he read the letter. "Why should I go to go to the school and talk about you? And why do they want you to come with me! What mischief have you been doing?"

"Father, I don't know!" KiemGiok replied, thoroughly alarmed.

"Have you been smoking, or stealing, or something like that?"

"No Father! I've never smoked at school! You know I sometimes smoked one of your *kretek* cigarettes, but only when you gave me permission, and anyway, it made me cough! Oh, sometimes Budi's father gave us a puff, but we never took his cigarettes. And I have never stolen anything that I knew belonged to someone else. I have no idea what the teachers want to talk to you about. You know my last report was good."

"Have you been going to more cock fights?" demanded his father, eyeing his son carefully.

"You know Mother punished me severely for that," KiemGiok replied truthfully, but evasively.

"True," replied Father, calming down somewhat. "Well, we will go, and see what they want."

The following day father and son, in pensive moods, walked from their home on *Jalan Setasiun* to the school on *Jalan Bitingan*. Both were utterly astonished by the reason for the conference and the proposition the head teacher made.

"Thank you for coming," Headmaster Chen began. "The teachers have noticed KiemGiok's consistently superior work, and his willingness to help others. We have discussed what middle school we should recommend when he finishes primary school in a few weeks. We strongly recommend that he attend the Chinese English School in Semarang."

After a stunned silence Father remembered his manners sufficiently to murmur, "Thank you for your kindness." Further words failed him.

Semarang was an important Dutch seaport, one of the great cities of Indonesia. But Indonesian freedom fighters and Dutch colonialists were still battling it out in Semarang, and wise people in Kudus avoided going there. Clearly, this teacher believed he was suggesting an honor for KiemGiok, but how could he possibly afford to send his boy there, Father thought, even if it were safe? He and his wife worked hard to support their family. He had generously allowed his sons to stay at school, unlike his daughter who joined the family workforce as soon as she could read. KiemGiok had been given a good opportunity to learn. He had already spoken to his employers and was proud to obtain the promise that his son could work as a clerk in the factory office as soon as he left school.

But Headmaster Chen had not finished. "We realize the school we are recommending is very famous and rather expensive. But all the teachers here think KiemGiok has brought honor to our Dutch school, and we have contributed to a scholarship to pay his fees at the Chinese English School."

This time Father was too shocked to even say a polite thank you. KiemGiok gathered his wits and said, "Thank you Mas (teacher) Chen. But I have nowhere to stay in Semarang."

"That is a difficulty," the head teacher replied, nodding sagely. "Do you have family or a family friend you could stay with?"

Sie TjwamKhing found his voice. "Thank you for this honor," he said firmly. "But I do not have the means to send the boy to this school, even if his fees are paid. This simply cannot happen." He rose to go.

Headmaster Chen was not easily dismissed. "Of course you will need time to arrange everything," he smiled disarmingly. "Please go home and let me know when you have organized it all. We will meet in three days' time."

Father Sie opened his mouth to protest, but KiemGiok was faster. "Yes, sir! My father and I will talk things over with my mother and let you know when everything is arranged."

Headmaster Chen stood, indicating the interview was over, and father and son were ushered outside.

"What did you know about that!" exploded Father Sie as soon as they were beyond the school gates. "How dare you agree we would talk about this and arrange everything! It is impossible! I say, impossible! You cannot go!"

"No Father!" Excitement made KiemGiok's voice crack and squeak. "No father, no. I did not know what Mas Chen was going to talk to you about. But please father, please, this will give me the chance to learn about science! Maybe I can become like Mas Einstein himself! Maybe I can discover the secret of the sounds of the *erhu*! Father, you cannot say no to me!"

"Ridiculous! Absolutely ridiculous! We can never afford for you to go to this fancy school!"

"But didn't you hear, father? The teachers will pay! They can afford!"

"How can I allow such charity? How dare you suggest such a thing? Our family has never begged! It is preposterous!"

"But this isn't begging! You heard what he said. It's a scholarship!"

"And didn't you hear that he said 'all the teachers have contributed?' I repeat: we are not beggars and we do not take charity from anyone. We don't need those teachers to take up a collection for you! I trained you to be independent, not a beggar!"

KiemGiok knew it was unwise to talk to his father any more. But this offer was the chance of his life. His only hope was that his mother might approve. They walked the rest of the way home in tense silence. Once home,

KiemGiok busied himself with chores: feeding hens already fed, checking and rechecking that the overflowing water bucket was full, sweeping and re-sweeping the yard. His father sat staring at nothing, and was still sitting in silence glaring through the open door when his wife arrived home.

"What did the teacher want? What had KiemGiok done that was wrong?" she demanded, sensing the tense mood of her husband and son.

Silence.

KiemGiok could feel his mother's eyes boring into him, searching for his misdemeanors.

Father spoke. "Relax woman, he hasn't done anything wrong."

"Then why are you both so . . ." she began.

"It's simply out of the question, that's all. We can't do what the teacher asked. The head teacher, Chen XiJiong, wants KiemGiok to go to the Chinese English School in Semarang. He says the teachers here will pay his fees."

"Oh, KiemGiok! That's so wonderful!" His mother clasped her hands joyfully.

"Don't be ridiculous woman! How can we possibly accept such charity? Anyway, I have promised he will start work in my office as soon as he finishes school this year."

"You haven't said no, have you?" gasped his wife.

"No Mother, we haven't," said KiemGiok earnestly, immensely encouraged by her support. "We go back and talk to Mas Chen in three days when we have arranged everything. Please, we can work it out, can't we? There must be some way for me to go to this school! I just need to find somewhere to live!"

"Impossible! Impossible!" muttered Father Sie repeatedly to himself. His wife ignored him.

"They'll pay all your fees?" she asked. "All your fees?"

"Yes."

"But you have to find somewhere to live, right?"

"Yes."

"Then you can stay with my brother, Tjan IkKhoen, who lives in *Kampung Sukolilo*. I'm sure he will allow it if you help with chores."

Chores! "But who will look after my hens?" KiemGiok suddenly blurted out.

"Oh, KiemSiang can do that," said his mother nonchalantly. "He's old enough now. It's time he took some responsibility."

"Woman, you should be getting the meal, not interfering," said Father Sie with the heavy authority of a man who knows he is defeated. His wife bustled off to prepare the peanut sauce for *gado gado* and KiemGiok decided to play a little *erhu* music. With his father so grouchy he did not dare ask his mother if she had any of his favorite *krupuk* crackers to top the *gado gado*.

"Cut that noise!" said his father irritably, as KiemGiok hit the third note.

KiemGiok silently laid his *erhu* aside, and furtively slunk out to the chickens scratching under the banana trees. One of them, clucking shrilly, had just laid a very large egg. Triumphantly he presented this to his mother. Under the influence of her twinkling eyes and beaming smile he relaxed.

Next morning mother left her shop in KiemLiang's care, and took the first train to Semarang. She returned mid afternoon and announced triumphantly at the evening meal that all was arranged. Her brother would be delighted to host KiemGiok in exchange for unspecified chores. Exactly three days later Father Sie sent a letter to Headmaster Chen. It politely declared that KiemGiok's lodging had been arranged and he could proceed with the plan to go to the Chinese English School in Semarang.

*****

Oh, the bustle and the hustle! KiemGiok stood at his front door surrounded by an untidy pile of goods. The family scurried nervously around. Mother presided over a small bag of clothing, muttering, "He needs more shirts" (or shoes or whatever). Father kept counting boxes and bike and bags, till KiemGiok earnestly wished he would go somewhere else and do something else.

KiemGiok had no interest in clothes, but he checked and rechecked that his *erhu* and books were in the box he had made for them. Then he grinned to himself. Just like father!

At that moment KiemSiang came running into the house shouting "Taxi! Taxi!"

The family chose to spend money on a taxi because the traditional horse and cart was slow, and frankly uncomfortable, and trains were becoming very unpredictable. In fact, before the year 1949 ended all trains between Kudus and Semarang stopped. With a truce between the Indonesian freedom fighters and the Dutch only recently declared, KiemGiok's parents agreed a taxi was the most secure way for their son to travel. The taxi was a very small, very ancient, two-door sedan, only just functional.

Mother carried the precious bundle of clothes. Father picked up the ancient bike found in a garbage pit and painstakingly restored by his son to function, and strode through the gate. KiemGiok struggled with his boxes of books and *erhu*, and eventually gave in and made two trips. KiemLiang appeared cuddling her kitten, self-consciously carrying her parting gift for her brother: a set of water color paints. KiemSiang sauntered out proudly bearing his mother's gift held behind his back, a box of wood-working tools. He knew his big brother was always whittling away on a stick with a knife, and had suggested to mother that proper tools should be fun for him. Unobtrusively he slid the toolbox into a book-box while Father and the taxi driver tried to secure the bike into the already crammed trunk. Eventually they admitted defeat, and the bike was tied precariously to the roof of the car. Father arranged himself majestically beside the driver's seat, and KiemGiok squeezed into the tiny space left on the back seat. It was the first time either father or son had ridden in a car.

When the loaded taxi reached the corner of the street it was suddenly hailed by an urgently waving boy.

"Budi!" shouted, KiemGiok, embarrassed by a sudden prickling behind his eyes. Budi would not be going away to school. He was very happy to be offered the clerking job intended for KiemGiok at Father Sie's *kretek* factory.

"Here, man," Budi thrust a package through the car's open window. "I know you like noise! Father made this for you."

KiemGiok's eyes misted over. "Thanks man!" he said, and with that Budi was off. Later, when he opened the parcel, KiemGiok found it contained a small wood carving of *a bonang* set, beautifully made. On its base was written something in a script KiemGiok could not read. He suspected, and later found was true, that the little Arabic-inscribed gift had been taken to the Menara Mosque and blessed for him.

Soon his eyes were glued to the window as they travelled the eighty kilometers to Semarang. At first the countryside was much the same as the villages where he roamed with Budi and the buffalo, but as they neared the city the crowding astonished him. When they passed a large and most beautiful Chinese-style temple his father asked the driver, "What's that?"

"Oh, that is the Gedung Batu Temple, or you call it the Sam Poo Kong Temple," relied the Indonesian driver. "Do you know its history?"

"Ah, yes, I thought so! Of course I know its history," answered Sie TjwanKhing proudly. "It was founded by the Chinese explorer Sampo Taikong, whom we call Zheng He, about the year 1410. He and some of his crew are buried there."

"He was a good Muslim you know," said the driver, pompously.

"Yes, he was. He loved all people, and allowed all to worship peacefully."

"Yes, it is still a place of prayer for Muslims, and even for you Chinese and Buddhists."

"But it wasn't always like that, you know." Father Sie said in disgust.

"True," said the driver, "A hundred years ago a Dutch man bought it and actually charged people to pray! Charged people to pray, did you hear! How bad can you get!"

"Terrible!" Father agreed, adding proudly, "Ah, but then the great Chinese sugar merchant Oei TjieSien bought it back and repaired it, and people came back to pray. Now it is a good place for both Chinese and Indonesians."

"That sugar merchant!" responded the driver disdainfully, "He traded anything, even opium! We call him man of two hundred million! He was made of money. He made his money out of us, our products, our country and we should have his property!" The taxi driver spat out the window and blasted his horn defiantly.

His passengers greeted this outburst with silence, and, quickly mollified, the driver hastened to find another landmark to point out to them. Although they could not help being Chinese, they did speak Indonesian perfectly, and they were obviously proud of their country.

"Would you like me to take you past the Blenduk Church? It's the oldest Christian church here, and is about two hundred years old, but of course not nearly as old as the Gedung Batu Temple. I won't charge you extra, because you've already come a long way, and it's market day so I can easily get a fare back to Kudus."

"Why not?" answered Father, thinking, "I might as well make the most of this trip."

KiemGiok enjoyed the sightseeing, but was anxious to get to his uncle's home and organize his life there. School would begin the following day, and he wanted to be ready. But his gratitude at being allowed to attend the Chinese English School restrained his irritation, and he said nothing. However, once they had seen the church, Father also lost interest in sightseeing, and directed they be taken immediately to *Kampong Sukolilo*. The driver grunted and swung his wheel so hard they grabbed their seats in fright, and in silence wove through increasingly crowded city streets.

KiemGiok tried hard not to gasp when the taxi screeched to a halt and the driver announced, "We're here". The *kampong* (village) was densely crowded, so unlike the leafy village he was expecting. His uncle's house was much smaller than his own home, and crowded between other houses. There was no yard, no banana tree, no well.

Father Sie climbed out of the taxi, and before he had time to knock a pleasant-looking woman, his sister-in-law, greeted them. She invited them in for tea, but Father indicated they must pay the taxi, and bring KiemGiok's belongings inside first. As the bike and boxes came out of the taxi, aunt stopped smiling.

"He's got a lot of stuff! I don't know where to put it!" she complained, scowling.

Father Sie seemed undisturbed by this news, but KiemGiok noted its ominous implications. His precious belongings were dumped unceremoniously in the street at the front of the small house, and the taxi disappeared. Father would return by bus, to save money. While father and aunt sipped tea and gossiped about family, KiemGiok sat on the edge of his stool, longing to bring his treasures to safety. Eventually Father took his leave, amidst many admonitions for good behavior. Then, with assurances to his sister-in-law that if KiemGiok proved a problem he could be returned to the bosom of his family at any time, he departed for the bus.

"You can't keep all that stuff here!" Aunt declared belligerently as soon as her brother-in-law was out of sight and hearing. "What on earth is it, anyway?"

"It's my books for school, and . . . and . . . other things," stammered KiemGiok. "Didn't mother tell you I was coming here to school?"

"Oh, she muttered something like that, but she never indicated you would have all that stuff. She did say you would be willing to help with chores!"

"Of course," replied KiemGiok.

He grabbed his small bag of clothes so lovingly prepared by his mother and hastily followed his aunt inside. "You'll sleep there," she said, indicating a mattress propped against the wall beside the entrance door. "You can see there is no room here for your stuff."

After his aunt and uncle went to bed KiemGiok dragged his things inside, and piled them around his mattress. He slept fitfully.

*****

Next morning KiemGiok was a strange sight as he peddled his ancient bike to Semarang's famous Chinese English School. With an *erhu* tied to his back, the larger of his book boxes precariously balanced on the bar of his bike, the gifts from his family and Budi and the small box of books protruding from the basket hanging between the handlebars, he was almost hidden from view. He didn't so much dismount as fall off his bike, clutching the handlebars determinedly. Reassuringly, other boys were arriving on bicycles as ancient as his, but none so heavily laden.

"Where do I keep my school books?" he asked a passing teacher. "I'm Sie KiemGiok, from Kudus."

"Ah, so you're the boy from Kudus," the teacher smiled, noting the new boy's anxious face. "Students don't usually keep so many books at school," he said. Appraising the boxes perched on the bike, the *erhu*, and the apprehension in KiemGiok's face, he added, "I'll see what can be done."

And thus KiemGiok was given a corner of the classroom for his books and *erhu*. From the beginning, the school made him welcome, and gave him every opportunity.

At the end of the first day, KiemGiok took two of his books home to complete assignments. Aunt was waiting for him.

"Today you will come with me and I will show you where to buy food from the market. After that you can clean the floors, and then Uncle wants you to help him repair the roof."

That night KiemGiok sank exhausted to his mattress on the floor, his two unopened books lying beside him. He was too tired even to think of a protest. It took only a few days for him to learn to ask his aunt before he left for school for her list of requirements from the market, to expedite that chore efficiently on his way home. It took him longer to realize he could never do school assignments at his uncle's house, so he did them at school, during recess or staying after lunch when other students went home. Although school began early, it finished by midday, and there was plenty of time in the afternoon before he needed to get food for his aunt for the evening meal. She made no protest when he told her he was staying at school to complete assignments, simply stipulating, "as long as you have my supplies for the evening meal and do your jobs after that." Often when he was working on assignments teachers would stop and talk, sometimes helping if there was something he wanted to learn. He got to know them all very well.

Aunt had an endless list of chores, and she filled every moment of his time once he arrived from school. Despite frustration at having little time to play his *erhu* there was one chore he really enjoyed. Aunt's collection of cooking pots was ancient and battered, and she regularly asked KiemGiok to take one to a nearby mender.

KiemGiok was entranced by the pot mender's ability to turn his work into music. As he hammered the battered and broken metal pots into shape, he made cheerful rhythms, and a symphony of tinkling sounds. He sounded exactly like a one-man *gamelan*! Whenever he had opportunity KiemGiok lingered at the mender's house. Fortunately, the man did not mind a teenage boy watching him work.

One day KiemGiok brought a pot to the mender's house later than usual. After hard persistent knocking the mender opened the door, and KiemGiok was surprised to see he was splendidly dressed in a blue batik jacket with a gold embroidered *songkok* on his head. The usually friendly mender frowned and shook his head sternly.

"No, I cannot mend that pot today. I have an appointment. Your aunt can wait."

KiemGiok looked at the wonderfully transformed man. "Please, Mas Toha, where are you going? You look splendid."

The pot mender smiled. He was not used to being called Mas but he liked it. "I play *gamelan*, every night," he said. "I play a guitar."

"Oh Mas! How wonderful! Please, please will you teach me?" KiemGiok's shining eyes were utterly persuasive.

"Oh, I don't play like a master. I just know how to make a good tune with my hands."

"But you will show me how?" pleaded the boy. "Please! I love music!"

"Come back tomorrow," said Mas Toha. He took the broken pot and abruptly turned into his house. KiemGiok did not see the tears in his eyes, or know the sudden surge of joy the man felt at this unexpected opportunity to share his music.

After this encounter, KiemGiok tried to visit Mas Toha every day on his way home from school, and Aunt's broken pots always took several days to mend. She grumbled, but never realized why. KiemGiok discovered that Mas Toha could read only very basic Indonesian, and understood just enough of numbers to be able to charge his clients. He had never been to school. He learned to play guitar simply by copying his father, but his musical talent and memory were extraordinary. After hearing any piece of music just once he could play it, as well as making beautiful and complicated variations. Sometimes he skilfully imitated three or four different instruments together using just his guitar. Eagerly KiemGiok soaked up everything his teacher told or showed him about guitar music.

KiemGiok's uncle, Tjan IkKhoen, was more friendly than his wife. He knew his nephew had very little money, and made useful suggestions to improve the teenager's financial situation. He introduced him to a newspaper delivery round. This made good use of the ancient bike, and brought valuable money to pay for school extras like writing paper. When he noticed his nephew not only drew neat pictures in the margins of unsold newspapers, but colored them attractively with the paints KiemLiang had given him, Pak Tjan mentioned his nephew's skill to a friend who ran a photographic shop. Soon KiemGiok was spending two hours two afternoons a week tinting black and white portraits of the shop's wealthy clients. It was delicate and exacting work, but KiemGiok enjoyed it. He was grateful for his increased financial independence, and appreciative of opportunities to improve his drawing skills.

KiemGiok's habit of staying after school not only meant teachers came to know him well, but an increasing number of classmates stayed to ask for his help with their work, especially math and science projects. This took KiemGiok's time, but he discovered explaining to others helped him to learn better. When the reports from the year-end exams came in he had scored a perfect one hundred per cent in all subjects.

Towards the end of his second year at the Chinese English School, as he was gathering his books ready to take to his uncle's home, hoping he would be allowed to keep them there, head teacher Koo TjinToen approached him.

"KiemGiok, I would like to talk to you before you leave."

"Yes, Mas Koo," replied KiemGiok anxiously.

"You have done well in your studies."

"Yes, Mas."

"You spend time at school after classes. Why?"

"It is just good to study here, better than where I live."

"I see. Like that is it?" KiemGiok nodded. "We thought so. Well, I have noticed how well you are doing, and wondered if you would like to live here at school next year. You know that small room beside the physics laboratory? That could be yours. You could sleep in there. There would be room for your books there too."

"But, Mas Koo, I don't . . ."

"Yes, yes, we understand you don't have any money. But that is not a problem. You could do some work for the school, like some typing and copying some papers in the office and guarding the school at night. You could help the students who are not doing so well, and that would pay for your lodging and your meals at the school canteen. What do you think? Could you manage that?"

KiemGiok simply stood, eyes shining brightly, hardly daring to breathe lest what he heard became a vanishing dream. "Oh, Mas!" he exclaimed at last. "Would I really be allowed to do that?"

"We would all be very pleased if you would agree to it. Come with me. No, wait. Pack your things, and I'll write a letter to your parents. I'm sure they will agree, and you will be a real help to the school."

A few days later when KiemGiok swung out of the school grounds on his old bike he had only one small box of books precariously balanced on the handlebars. The rest of his belongings were locked into the small room beside the laboratory, and the key was pinned in his pocket. He was puzzled that the room had a very nice new mattress and bed in it, which he had not expected, and a neat table and chair for him to use as well. He had always thought it was just a storage room.

He was so excited about his new prospects that he decided to take the bus home immediately. He helped aunt with her shopping needs, then told her he was going home that evening, because his parents needed him. He had been saving money to buy a second hand guitar (the old battered one loaned him by Mas Toha was literally falling to pieces and efforts to repair it had been futile), but on sudden impulse he decided to treat himself with a taxi, and share his good news with his parents as soon as possible.

The family were eating the last of their evening meal when he walked through the door. Mother uttered a cry of delight, threw her arms around him, and the bananas she was carrying scattered across the floor. KiemLiang and KiemSiang jumped up, clattering their stools and shouting, "KiemGiok! KiemGiok! How did you get here!"

Sie TjwanKhing, dignified as ever, remained seated. "Welcome home son," he said.

KiemGiok grabbed KiemSiang's overturned stool, and sat down. "I'm home!" he announced. "I've got good news! Oh, and would you like to see my school report?"

He pulled the envelope from his trouser pocket, and even Father Sie got up to look. There was no need to scrutinize the report. The long, unbroken row of multiple one hundred per cents spoke for itself.

"Proud of you, son," Father said simply. "You have done well again."

"Very," added his mother.

"Well done bro!" grinned KiemSiang, and slapped his brother's shoulder.

"How long are you home this year?" asked his mother.

"Eight weeks," answered KiemGiok, "but I want to work. My teachers have given me assignments so I can make a start and be ready for next year."

"Good," said his father. "No sense in just fooling around."

"Aw, Dad," said KiemLiang. "Work is all you ever think about!"

"And I have something else to tell you," said KiemGiok, pulling Mas Koo's letter from his pocket. "I don't need to go back to live with Uncle Tjan. The teachers say I can stay at the school."

"But . . ." began Father Sie.

"Don't worry, Father. You won't have to pay any money. The teachers said I can work for the school and that will cover my living expenses. Here, read what Head Teacher Koo says."

"What sort of work?" asked his father, looking at him sharply through slitted eyes.

"Typing and office work, like you taught me, and teaching other students and . . ."

Father took the proffered letter and glanced through it. "That's fine," he answered without further questioning. "You can can stay there. I'll talk to Ik Khoen." KiemGiok noticed his father looked very pleased.

Business over, the family settled down to chatter long into the warm tropical night. The last thing Father said before he went to bed was, "There's going to be a good *kroncong* performance in the *aloon aloon* central park tomorrow. We'll go there, son."

Next day KiemGiok strode proudly beside his father as they walked to the park. He was now as tall as his father, and felt strong and manly. The hardships from years of war had sprinkled grey in his father's hair, and his shoulders drooped. Somehow their roles seemed almost reversed from the time they made the first visit to *aloon aloon* park. As they walked KiemGiok determined he would never be a burden to his parents, never, ever, and he would take full responsibility for his life.

"Father," he ventured as they stood waiting for the musicians to assemble. "Father, would it be possible for me to do some work at your office while I'm home? I don't need to get money for it. But if I learn more office procedures I can be really useful for the school, and it would be good. If I do this school job well, you won't need to worry about paying for my living expenses ever again."

"Yes, I can arrange that," Father answered thoughtfully.

Suddenly KiemGiok was startled by a wonderful sound. Like all but very rich Indonesians, the Sie family had no radio, but they often listened to publically broadcast news and music. A radio broadcasting in the park was playing traditional melodies over the gathering crowd, but now it played something KiemGiok had never heard before.

"Father, what makes such beautiful music?" he asked. "I've never heard anything like it!"

"Oh, that's a violin. It's a most mysterious and wonderful instrument, I think perhaps the best in the world."

The two men listened intently. As the last enchanted notes died away, the announcer declared, "That, ladies and gentlemen, was Fritz Kreisler, playing Antonin Dvorak's *Humeresque* for your enjoyment. Give him a clap!"

"Ah, Kreisler," sighed Father Sie, "he is the best."

"Someday I want to play a violin like that," thought KiemGiok, as the *kroncong* performance began. Later, he was shocked to discover that the only thing he could remember from the afternoon's musical performance was the Kreisler violin piece.

"Father," he asked as they walked home, "was that a special violin that Mas Kreisler was playing? I've heard violins, but they didn't sound that good!"

"Yes, son, it was. It is very old, hundreds of years old, and made by a man called Stradivarius in Italy. People say it is one of the Seven Wonders of the World, like the pyramids of Egypt, the Great Wall of China, that special white marble temple in India, and skyscrapers in New York. That violin is very expensive, and people say the art of making such violins has been lost, and no one can make them like that any more."

"That's very sad Father. Surely someone can learn how to make those violins?"

"It seems not, my son. Of course men still make violins, but they say his secret for making such good violins is lost. I'm glad we can still hear those great violins being played. Wonderful music!"

"Yes, wonderful." KiemGiok never forgot that conversation. I will discover that secret, he promised himself, even if it takes all my life. I will discover how beautiful violins are made.

*****

When the new school year began KiemGiok took a taxi to Semarang. He was delighted to direct the driver straight to the Chinese English School on busy *Jalan Bojong*.

"You're lucky to be going to that school," commented the driver, eyeing KiemGiok's simple clothes. "Do you have a rich uncle?"

"No, I work for the school and study there as well. I'm the school umm ... caretaker," replied KiemGiok, hesitantly but proudly.

"You look pretty young for that job," the driver observed disdainfully, "but good luck kid." Sixteen-year-old KiemGiok ignored the aspersion.

When the taxi reached the school he jumped out quickly, paid his fare, and pulled his things from the trunk. Irritated by the driver's comments, he was determined not to show it. Maybe he did look young, and maybe he was not rich either, but his teachers trusted him, and he would do everything he could to honor that trust.

Mas Koo was waiting for him at the school door. He looked at his watch. "Aha! You said you'd be here at three o'clock, and it's five minutes to three. Great! You are reliable. Put your things in your room, then come with me while I show you what you need to know."

The next hour was busy, as KiemGiok took his instructions. At last Mas Koo said, "You can get *gado gado* or *nasi* at the school canteen, and the school will give you an allowance for breakfast and dinner. But tonight come to my home. My wife is expecting you, and you should know where I live in case of trouble."

"But Mas, my mother prepared my food for tonight. I do not want to trouble you."

"It's no trouble, just an order!" Both men grinned, and soon KiemGiok followed the head teacher to his home. His mother's food would keep for breakfast, and he enjoyed the teacher's delicious meal, although somewhat surprised by the simple Indonesian-style home where he lived.

Later that evening, alone in the school, KiemGiok suddenly felt very young and very alone. Just in case, he would be prepared for trouble. His family and friends never locked their doors, and no one was ever robbed. But this was bustling Semarang, not pious Kudus. With the help of the streetlight he found a sturdy Y-branched stick, and using the rubber band that tied his clothing bag, he constructed a serviceable catapult. He spent a few minutes with some round pebbles and a piece of bamboo practicing his aim, and then settled down with his books. He tucked the shanghai under his mattress. He was the school caretaker for the next three years, and never needed to use that shanghai, but it was good to know it was there under the mattress, ready if needed.

The physics teacher, The TiangHin, was Holland educated, and his enthusiasm infected his pupils. KiemGiok was delighted to discover his holiday studies put him far ahead of the others in his class. Mas The suggested that if KiemGiok did a small amount of extra homework for a few days, he could join the next class. A similar surprise awaited him in math class. The math teacher, Chen ShiPei, an engineer by profession, made real and practical the problems he gave his pupils. The only classes KiemGiok found he was not ahead in were English and Chinese. He excelled in Indonesian and Dutch language, so decided he would just have to work a bit harder to make the strange foreign languages come alive in his head. Under normal circumstances, KiemGiok's moving to more senior classes may have caused jealousy and resentment in his previous, or new, classmates. He solved the problem by asking for his classmates' help with English and Chinese, and offering his help with math and the sciences. However, unexpectedly, one skill did cause jealousy in several of his friends: he entered the school drawing competitions and won several prizes. Although completely self taught, his work coloring photographs had given him opportunity to observe both good composition and coloring techniques.

Work in the school office was not onerous, taking only one or two hours a day. The experience gained at his father's office proved invaluable, and he did his work quickly, much to the satisfaction and appreciation of his teachers. He could now use the time spent biking the three kilometers to and from his uncle's home for study, and still have occasional opportunity for visiting. He did not neglect his aunt and uncle, and they always welcomed him and offered food. Sometimes he did small chores, especially offering to take broken pots to the mender. Aunt always expressed gratitude for any help, and he realized she missed the support he had given her, and even his company, as her children had all long since left home. He never mentioned the main attraction for visiting was the pot mender. He continued learning guitar from Mas Toha, playing from memory attractive Indonesian folk music. Mas Toha helped him find a better guitar, and by spending carefully on food KiemGiok was able to buy some quality music books: Ferdinand Carulli's classic guitar school, and Mateo Carcasi's etudes for guitar. He learned and practiced guitar from these masterpiece books for many, many years after he left the Chinese English School.

*****

Spending much of his time at school gave KiemGiok an unexpected opportunity. Previously music had been a personal interest which his father encouraged as a hobby. Now, as he performed caretaker and office duties he became aware that various groups at school practiced for public performances. Mas Koo encouraged his pupils to learn a variety of skills, and public concerts advertised the school. KiemGiok enjoyed his classmates' activities, and soon they invited him to participate. His guitar skills were in demand to accompany other performers. He did not aspire to star performance, content to offer back-up skills to others. But he did enjoy acting, especially magic tricks. He became particularly adept at playing ping pong, winning several medals and two grand cups. On weekends he explored the crowded, bustling streets of Semarang, and became good at playing billiards.

One day for a school assembly performance he was asked to accompany a classmate who played violin, and readily agreed. It was a lively piece, fun for both young musicians. Later on impulse, KiemGiok asked his friend if he could play his violin. "Of course!"

Gingerly KiemGiok picked up this mystical instrument, stroked the bow across the strings, only to make a disconcerting cat-like screeching. "Press the bow down on the strings harder," advised his friend, chuckling. Suddenly the room was filled with his, KiemGiok's, magical violin sounds. He pressed the fingers of his left hand against the strings, and the sounds changed. He tried playing his guitar accompaniment on the violin, and found he could. He grinned playfully at his friend.

"I'm busy tonight," his friend remarked. "Why don't you take the violin and have fun on your own?'

"Ah, thanks. Are you sure?"

"Yes, yes. Actually, it isn't my violin. It belongs to the school. I play the clarinet in the school orchestra, and just do this violin thing for fun. I'm sure they won't mind."

KiemGiok stroked the violin. It might not be a Stradivarius, but it was beautiful, smooth, and shiny. "Thanks a lot," he said.

That evening physics and math assignments were untouched. Time raced, and it was midnight before he turned off his light. But by then he had worked out the basic harmonic patterns of the violin, and how to play Indonesian folk tunes that Mas Toha had taught him.

When he returned to Kudus the next school holiday he asked his father if he could get a violin. A few days later Father returned with an ancient, out of tune, and badly scratched violin. "My friend said you can have this," he reported.

"For me?" exclaimed KiemGiok. "Really, for me?"

"My friend wasn't sure if it is usable."

KiemGiok rubbed the scratched surface down with several applications of wood polish and transformed its appearance. His first attempt at tuning broke the e-string, but with patience he soon had a usable instrument. He spent hours studying its structure, trying to work out how the sounds were produced.

*****

KiemGiok looked up sharply at the gentle knock. His school office duties were finished for the day, and he had settled down to complete his assignments. He was a well-organized student, with a well-planned timetable. Rather than learning by rote as did most of his classmates, he liked to master the principles of each subject. Although he was generous with his willingness to help his friends, he regarded the time after his evening meal as his own, guarding it jealously from interruption. So, who could be so inconsiderate?

Words of censure died on his lips when he saw it was Ong BingGian. BingGian was a serious, dedicated student who never asked for his help, although often he wished she would! Not only was she a good student: she was beautiful and kind. He wanted to get to know her better, but had never been able to think of any appropriate reason. Now there she was, standing at his door when all the other students and teachers had left the school, and of course he could not invite her in!

"Yes?" he said, horrified at how brusque he sounded. "Can I help you?" he added, which sounded scarcely any better.

BingGian cleared her throat, and fiddled nervously with a pen.

"We are having a celebration at my home for my father, Ong HokKwie, next Friday evening. My mother would like me to sing for the occasion. Would you, please, would you, I mean, would you be willing to accompany me on your guitar?"

KiemGiok stared at his classmate. Had he heard right? Was he dreaming, was this for real? It was unbelievable that he should be given this opportunity! He shook himself to make sure he was truly awake, and BingGian misunderstood his gesture.

"Oh, I am sorry, I did not mean to bother you. Please forgive me. I will find someone else to help me." And she turned to go.

KiemGiok collected his wits quickly. "No! No!" he almost shouted, and BingGian stopped, frightened at his outburst. "No! No! No! I was just thinking how I could help you, and yes, yes, I think I can. Yes, um, yes, I would be willing to accompany you next Friday. But I'm not that good. Um, er, um, shall we practice immediately after classes tomorrow to see if I am good enough?"

Suddenly BingGian smiled, a joyous radiant smile. "Oh thank you KiemGiok! That is so good. Yes, we can practice after classes tomorrow. I'm sure one practice will be enough."

She turned, and fled through the school gates, as though afraid he would change his mind.

KiemGiok's well-organized study plan for the night was shattered. It was fully half an hour before he could settle to physics formulae. He had accompanied so many people that no one noticed next day when he and BingGian walked to the school auditorium and began their practice. She had a delicate voice that suited the traditional, sweet style of folk song she had chosen. Most of the songs KiemGiok already knew from Mas Toha, so the rehearsal was easy.

When they finished BingGian handed him a piece of paper describing the address of her family, complete with detailed directions, then she ran through the school gate without a backward glance at her accompanist.

KiemGiok walked slowly and thoughtfully to the school office, and was almost there before he remembered he had not had any lunch. Fortunately, the canteen was still open, although there was very little food in the servery.

On Friday night he dressed carefully. He had no formal clothes, but he made sure those he wore were spotlessly clean. He decided to walk rather than bike so he was less likely to arrive hot and sweaty. But he overestimated the distance, and arrived at the gates of the elegant traditional Chinese-style house half an hour before the allotted time. He examined every tree and rock on the block in that half hour.

BingGian's mother welcomed him warmly when he knocked and gave his name, ushering him into a beautiful room. "There are pumpkin seeds and tea you can enjoy," she said, gesturing towards a gleaming lacquer table. Several older people were in the room, busy in conversation. KiemGiok picked up a few seeds, found a stool by the window, and sat watching other people arrive. There was no sign of BingGian.

The room filled with chatting and laughing people, and after what seemed like two years of waiting a couple of his girl classmates appeared. "Hello KiemGiok," they smiled. "What are you doing here?"

Their question annoyed him. "BingGian asked me to accompany her when she sings," he answered.

"Oh, that's right. You play guitar!" They turned, and walked away giggling, leaving him suddenly flustered beside the window.

After more hours of waiting, BingGiang's family joined the throng. Mr. Ong loudly cleared his throat, and the chattering died away. "Thank you for joining us here tonight. I do not deserve this little celebration, but the family wanted me to have one, and I know they have done this out of their esteem for me. I hope you have all found the tea and pumpkin seeds to your satisfaction. While we are waiting for the food to be served, we have prepared a few items for your musical enjoyment. First, Mas Toha has brought his *gamelan* group for your enjoyment."

KiemGiok gasped at the sight of his teacher. How had he not noticed him in the crowded room? As the group assembled, he realized he had never actually heard it play. It was a wonderful performance, and the guests clapped and shouted their approval when the last notes died away.

"Now I have asked my friend from the Semarang radio station to play for us. I'm sure you will enjoy him!"

To KiemGiok's surprise a European stepped forward, another person he had not noticed in the crowd. He reached under a table, and brought out, to KiemGiok's even greater surprise, a violin. In perfect Indonesian he explained he would play a medley of classical European and Indonesian music. KiemGiok listened entranced. He remembered the magical performance he had heard with his father in the *aloon aloon* park in Kudus and this performance seemed just as good. Suddenly, to his delight, he heard *Humoresque*, the piece he remembered so well.

After several more excellent performances, Mr. Ong said, "And I have kept the best for you till last. My daughter has agreed to sing for us, and she has a friend to accompany her." A friend! KiemGiok thrilled at the title. He strode confidently and purposely to the front of the room while the most beautiful vision in pink moved to join him.

He had made sure his guitar was tuned carefully, and he struck the first chord confidently. BingGian smiled nervously, and he played a few notes to help her relax. Then she nodded, and they began. Her sweet voice floated beautifully above the rippling guitar, and when they finished the guests roared their appreciation. Mr. Ong looked very pleased with himself. "My daughter only sings a little," he said, "and she is completely untrained, but of course, to a father it is very sweet! But now, it is time to eat, so please move to the tables and enjoy the food."

As guests walked to the food tables KiemGiok noticed Mas Toha and the European violinist both slipped quietly out the door. He had planned to do the same, but now felt he could not waste the opportunity to get to know BingGian. Just then BingGian herself came over, thanked him for his accompaniment, and invited him to join her at the family table. He smiled shyly, but needed no second invitation.

When the school learned that KiemGiok and BingGian made a good musical duet they were included in many of the school's public performances.

One day BingGian called at his room to arrange a practice between them, and found him playing his battered violin.

"I didn't know you could play the violin," she said. "How long have you been taking lessons?"

"Oh, I don't learn. I'm just teaching myself. A few years ago I heard Fritz Kreisler playing on the radio, and ever since I wanted to play."

"You should take lessons. You do very well."

"Yes, it would be good to learn properly. But who teaches violin around here?"

"I'll find out."

A few days later she met KiemGiok in the school hallway. "I talked to Father," she said, "and he says he will ask Anton Piontek if he is willing to teach you."

"Who's he? That doesn't sound like an Indonesian or Chinese name."

"Oh, he's a very good violinist. He works for the Semarang radio station. You may remember he played for Father's celebration. He's Dutch, although Father declares he's German, not really Dutch. They say he's in Indonesia because really he's a Jew, and he escaped from the Germans who were killing all the Jews."

KiemGiok frowned. "I cannot understand how people could do something like that; I mean, kill people just because of what they were born. But I guess Chinese people in this country are not always treated fairly. We Chinese are very hard working, but some people don't like those who have become rich. My family is not rich. We just want to be good citizens and help other people. But sometimes people have said very bad things to my mother in her shop."

"Yes, father has told me the same. But do you want me to talk to my father about Mr. Piontek?"

"Yes, of course."

*****

"Ah, I remember you! You're the young fellow who plays the guitar!" said Anton Piontek as he opened the door to KiemGiok's knock. "Now you want to learn the violin as well? Guitar not satisfying you, eh? A musician needs to be dedicated to his instrument, not flitting from one to the other!"

"Yes, Mas Piontek. I learned the guitar first. But then I heard a violin, played by Fritz Kreisler. I . . . I have taught myself a little about the violin, but I want to learn it properly."

"Well, come in, come in, don't just stand there, and let's hear you."

KiemGiok took his old violin from its case, and began to play.

"Hmm. So you haven't had any lessons?"

"No, Mas."

Anton Piontek sat down. "That's an awful violin this kid has," he thought, "but he shows real talent." Aloud he said, "Are you prepared to work hard? Really hard?"

"Yes, Mas."

"Pak Ong said you were in your last year of middle school, right?"

"No, next year will be my last year."

"And then?"

"I hope to go overseas to further my studies. I want to study physics."

"So you really want to study physics! And I'm supposed to teach you to play a violin in just eighteen months! Oh the arrogance of youth!"

KiemGiok was alarmed. "I will work hard, truly I will Mas."

"OK, I'll try. If you will follow me, maybe we can make it work. First, you need some good violin music. No fiddling around with folk songs like you've just played. Do you know the good music shops in town?"

"Yes, Mas, I've bought guitar music there."

"Is that how you learned to play the guitar?"

"Not exactly. I learn folk music from Mas Toha the *kroncong* player. But I got some books to help me learn classical guitar."

"I see. You do take music seriously. OK. Go to the music shop and buy the study program of Charles de Beriot; we'll begin with that. Learn each study carefully. Don't race through the book. If you want to be a *maestro* you must master every technique before you go racing on to something else. Do you hear? No rushing! I say, no rushing! Later, when you have mastered de Beriot we'll move to Kaiser, then you can do Kreutzer and Rode and Dont. But remember, don't race. Did you hear that? No rushing! Master everything step by step. And no playing songs or other pieces till you have learned the techniques properly. Did you get that?"

KiemGiok was busy writing down the names of these famous violin books. "Yes, Mas, I will do as you say, I promise. I won't rush!"

Anton Piontek pulled a copy of de Beriot from a shelf, and placed it on a music stand. "Let's begin," he said.

When the lesson was over and KiemGiok was putting his violin back in its case, Anton Piontek said, "By the way, how long have you had that violin?"

"Since last summer holiday. My father's friend in Kudus gave it to me. He used it when he was learning when he was a boy."

Piontek nodded. "Yeah, it looks like that," he said. "The violin is a very mysterious instrument. How it makes its music is not well understood, but we should use science to learn it secrets."

"Science is my best class. That's why I want to learn physics."

Anton Piontek raised an eyebrow quizzically. "Many people have tried to learn the secrets of the master luthiers. But someday, someone has to crack the secret. Give it a go, lad!"

Next day BingGian was waiting for him at the school gate. "How did it go?" she asked. "The violin lesson, I mean."

"Good."

"So you liked Mas Piontek?"

"Well, he's a bit abrupt. He scared me at first."

BingGian smiled. "He scares lots of people at first!"

"His house surprised me," said KiemGiok. "It's very simple, not European at all. There was an old European man there who didn't speak Indonesian, and it didn't even sound like Dutch to me. There was a young Indonesian woman servant. The house had no special furniture or anything. I thought all Europeans were rich!"

"Well, I don't think he has much money, but Father says he's a very good violinist. People say he came as a refugee to Indonesia. The old man is his father, and the young woman is not a servant. She's his wife."

"Really?"

Next week after KiemGiok played his assignments, Anton Piontek beamed with satisfaction. "You learned that well," he said. "I think we can move to the next study."

"Mas Piontek," KiemGiok ventured as he was packing up, "where did you learn violin?"

"My father," Piontek said, with icy finality in his voice. "Now, you know what you must do this week?" Clearly his teacher did not wish to discuss himself or his story.

One day, towards the middle of his last year at school, KiemGiok's math teacher stopped by his room when he was practicing his violin exercises.

"You enjoy playing, don't you?"

"Yes, Mas."

"You need a better instrument if you are going to play well. You will never get a good sound out of that violin."

"Yes, I know. I have tried to improve it, but it is not well made."

"Interesting. How have you tried to do that?"

"Well, my teacher does some violin repairs, and he showed me how to improve it, but without success. I've been learning woodwork myself too, so I tried to make it better according to what Anton Piontek told me."

"You've been learning woodwork? Really? How?"

"Oh, you know those Javanese[8] and Balinese[9] men who work under the trees down the road? The ones who make lovely sculptures? Well, for many months I've been going there in the afternoon to learn from them. They are happy to show me what they do and how they do it. They taught me how to cut wood, and let me make some simple sculptures. They insert pieces of shell into the wood to make patterns. I really want to learn to do that myself."

Chen ShiPei gazed at his pupil long and hard. "You're a remarkable kid," he said. "Whatever you do, you give it everything you've got."

"I try my best, Mas."

"Maybe I can help you," said Chen ShiPei, smiling. He strode off leaving KiemGiok puzzled.

Just before the year ended, very late one afternoon when everyone else had gone, classmate Chen GaiTai came to KiemGiok's room.

"For you," he smiled, and handed the astonished young man a violin case.

"For me?"

Cautiously KiemGiok flipped open the catches on the case, and turned back the lid. Inside was a beautiful, gleaming, German-made violin. He tucked it under his chin, and pulled the bow across the strings. A wonderful rich tone filled the room. Emotion flooded his mind and heart.

But when he looked up to thank his friend, he had gone.

---

8. Since Semarang is on the island of Java, these men were probably itinerant craftsmen.

9. Bali, another island in the Indonesian archipelago, but the only island that is Hindu not Muslim. These men had different artistic traditions that KiemGiok was interested in learning.

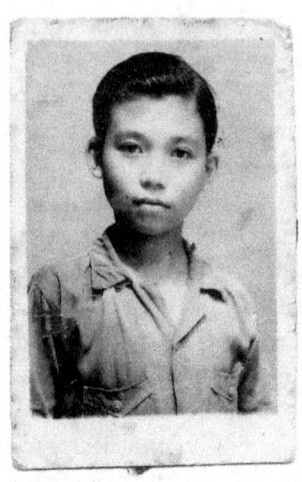

Sie KiemGiok in 1949, aged 14, when he started at the Semarang Chinese English School.

The Kudus Menara Mosque, with *bedug* tower L, and dome R

The *klengteng*, venue for Kudus *kroncong* and *gamelan* performances.

The Sie family home in Kudus, photo taken many years after KiemGiok left.

In this 1949 Kudus school photo KiemGiok is on the far L in the front row wearing darker clothes.

"Four Clowns" in a Chinese English School performance. KiemGiok is second from right.

KiemGiok doing card tricks in a school performance (note the signature *Sie KiemGiok 1953*).

Receiving the school ping pong championship cup.

KiemGiok, the guitar accompanist.

Anton Piontek.

Formal portrait, Sie family, 1955. Back row, L to R, KiemLiang, KiemGiok, and KiemSiang. Front row, father Sie TjwanKhing, L, and mother Tjan SingNio, R.

Second Movement

Chinese Pipa

Anton (alias KiemGiok) leaned against the rail of the *Tjiwangi* and peered into the looming shadows of early dawn. Maybe others could rest at this time, but as the ship cut through the grey gloom into the bustling port of Hong Kong he was far too excited for sleep. The further he sailed from home the more he wondered if the price of his commitment to education and independence was too high. Fortunately, his youthful love of adventure overcame his doubts, bolstered by the noble dream of "rebuilding China" which he shared with other students on the ship.

Feeling a tug on his sleeve, he turned to see his friend from the Chinese English School. Suddenly the sun breasted the eastern hills, flooding the harbor with brilliance.

"Hey, we thought Semarang was a busy port," said Huang JingAn, swinging his arms expansively "but look at this!"

"Semarang's a good port," responded Anton, defensively, "but its waters are too shallow for large ships. I'm sure China has even better harbors!"

"Of course!"

Hong Kong harbor seemed huge. Numerous large ships unloaded to small *sampans* called lighters; a stately white passenger liner was tied to the Tsim Sha Tsui wharf; battered and rusty vessels registered in various parts of the world were in harbor for repainting; busy local ferries ploughed in every direction to their destinations; and large, classic Chinese junks modernly motored sailed sedately through it all, an exciting spectacle. The friends gazed in awe.

"It's the 13th of July," declared Anton.

"Wow! It's exactly one month since we left Semarang! Then we were excited to go to Tanjung Priok Harbor in Jakarta," laughed JingAn. "And now this! 1955 is going to be fabulous!"

Anton nodded agreement, and stared silently across the sparkling harbor. For years he had known exactly what he wanted to do with his life:

study physics and music. He dreamed of being an Einstein or a Stradivarius, preferably both. The problem was how to achieve this goal. His teachers urged him to attend Leiden University, but enquiries soon revealed finance was a massive problem. He was determined not to burden to his parents. He would have asked for small amounts of help from them, but the cost of going to the famous Dutch university was too great.

"It was a tough decision," said Anton, flatly, after a pause. He now realized how little he knew of the country he was about to enter.

His friend turned sharply and stared at him. "You?" he said. "You? You were always the one so keen to leave!"

"Not leave, study," corrected Anton.

Anton's special friend Ong BingGian had already gone to the United States to study medicine. Her family could afford the costs involved. He and BingGian had become very close in their final year at school; so close, in fact, that BingGian's father offered Anton financial help for his studies. But the cost of education in the United States was enormous, and Anton recoiled from dependency. Eventually he accepted Mr. Ong's generous offer to pay for the journey to Canton, but refused further assistance. He confidently expected that once he finished his studies he could be reunited with BingGian, and repay her father. But in the Hong Kong dawn BingGian and her family seemed far away, and Anton suddenly felt very lonely.

"Yeah, I hope they keep their promise," said JingAn, suddenly sobered. "I hated having to sign that document."

"They" was the Chinese government. The big attraction of education in China was the new Chinese government offered free education for overseas Chinese students, with the generous provision that they were allowed to leave China whenever they chose. It seemed an ideal arrangement. The People's Republic of China was young and described as developing well. The civil war that had torn it apart for decades had finally been resolved. Reports filtering into the Indonesian Chinese community were glowing. It was therefore no surprise that on the *Tjiwangi* were about twenty Chinese students from various parts of Indonesia, including several from the Chinese English School in Semarang. There was just one problem, to which JingAn alluded.

The Indonesian government did not share the enthusiasm of its Chinese community for the new government of China. When Anton applied for a passport to travel to China he learned two things. First, he was not Sie KiemGiok, but Sie Anton. Born when Indonesia was a Dutch colony, his father was required to register his birth with a Dutch name acceptable to the governing authorities. His passport would be in his official name. He had known from childhood that he had a European name; a few of his teachers at the Dutch primary school occasionally called him Anton, the name under which he had been enrolled. But his family and all his friends called him KiemGiok. However, he now counted it an honor to be known by the same name as his revered violin teacher, Anton Piontek. He had made good progress learning the violin under Mr. Piontek's tuition, and safely on board the *Tjiwangi,* in two aluminum boxes of treasured possessions, were, along with the German violin, all the violin teaching books Mr. Piontek had suggested: de Beriot, Kaiser, Mazas, Kreuter, Rode and Dont.

It was the second discovery the passport application revealed that was painful, so painful that sixty years later Anton still does not like to talk about it: if he chose to go to China he must sign documents forever relinquishing his Indonesian citizenship, and the right to return to his country. His mother was deeply distressed by this, and begged him to reconsider his plans. But with the optimism of youth, he believed he could find a way around the difficulties, and not be separated from his parents and BingGian forever. After all, he pointed out, he was free to leave China whenever he chose. After his China studies he planned to go to the United States and there re-connect with both BingGian and his parents.

"I guess your mother was as upset as mine," said JingAn, staring at Anton's somber face. Anton nodded.

Leaving Kudus and Semarang was difficult. Not only was his mother distressed, but it was painful to leave people like Budi and the Halim family. Budi's education finished at primary level. Besides his work for the *kretek* factory, for which he was grateful, he was busy helping his parents on their small farm. No longer was he a boy having fun and frolic with the family water buffalo. Now he worked hard ploughing rice paddies with the patient animal that had given him and Anton so much fun. Budi was still addicted to cockfighting, but Anton had long lost interest in this forbidden pastime.

To Anton's delight Budi and the entire Halim family were at Semarang wharf to see him leave. Several teachers from the Chinese English School were also there, including Headmaster Koo, physics teacher The, as well as Mr. Ong. Anton had never seen his mother cry, not even during the dangerous, frightening years of Japanese occupation. It was therefore very painful to see her, and his sister, weeping so disconsolately as they said goodbye. He tried to hold his own emotions, and merely acted wooden. But Budi's presence helped them all. He joked and talked freely of when they would meet again. Anton dared not tell him about the documents he had signed.

JingAn broke the tense silence and brought Anton back to the present. "It was fun to see Jakarta, and Singapore, wasn't it? But that storm wasn't fun. How did you keep your food down?"

Yes, the pain of departure was softened by the bustle and excitement of adventure. What a thrill to sail along the north coast of Java, which none of the students had ever seen, and arrive in the capital city Jakarta. Semarang, especially its Chinatown, always seemed big and bustling compared with Kudus, but Jakarta was unbelievably vast. An official from the Consulate of New China in Jakarta met the ferry, and organized transport for intending China students. Despite his help, although the distance from the Jakarta wharf to the Consulate was not great, it took more than an hour in the congestion of people and traffic. There they waited in a long queue of China-bound students, all needing travel documents. The consulate administrators repeatedly stressed they must hand over their passports to officials when they arrived in China. Anton wondered briefly what would happen when he wanted to leave if he had no passport, but he quickly dismissed the thought. Finally, all was arranged, and the students were on their way to the *Tjiwangi*. The ship, primarily a merchant transport, did not offer luxurious accommodation. But to the students she looked a grand and mighty vessel. The group was filled with a spirit of camaraderie and the excitement of exploring the unknown.

The *Tjiwangi* docked briefly at Singapore but passengers were not allowed to disembark. They gazed at the elegant colonial houses near the waterfront, and tried to be content watching the hustle and bustle of loading cargo. Several students grumbled about the prohibition to go ashore, but Anton was satisfied. He had no money, and was not attracted by rumors of luxuries in Singapore shops.

"I used physics," said Anton, suddenly responding to JingAn's question about keeping his food down. "Sit near the center of gravity where movement is least, and your stomach won't feel so bad. That storm wasn't nice, but it didn't last long. And the rest of the trip was great." He well remembered at the height of the storm, as the ship pitched and rolled, rows of greenish-grey-faced students lining the deck rails feeding the fish.

"Oh, you and your physics," laughed JingAn good-naturedly. "You solve everything with it!"

"Of course! Anyway, we're almost there," said Anton. "Tonight we'll sleep in China! Wow!"

"Wow indeed!" replied JingAn, adding, as though he had only just realized this, "You're right!"

They were told the ship would dock briefly for them to disembark at a passenger wharf, before moving to unload its cargo, and from the wharf they would take a train to Canton (Guangzhou). Reluctantly they tore themselves from the rails, and went below to collect their baggage. Anton had his ancient bike, two aluminum boxes of books and tools, and a tiny bag containing two changes of clothes and a woolen shirt his mother had made. His scant money was sewn into a pocket in his belt.

Down in the stuffy and crowded cabin fellow students were beginning to stir. "Get up! Get up, lazy bones!" yelled JingAn. "The ship's about to dock!"

Oh, let them rush and bustle, Anton thought, as he made his second trip up the narrow gangway with his boxes and bike. At least he was ready.

When the ship slid into the wharf the cool sea breeze died. The summer heat of Hong Kong hit like a furnace. Anton struggled down the gangplank with his heavy boxes, the first passenger off the ship, drenched with sweat. He glanced at JingAn. Rivers ran down his face too.

As promised, a China representative was waiting, prominently holding a large placard. How grateful the students were for that placard! All of them spoke slow, halting, school-book Mandarin, but none could understand a word of the guttural Cantonese that blasted their ears. It seemed Hong Kong had so many people they all needed to shout to make themselves understood. Anton had never heard people talk so loudly. Was everyone having an argument? he wondered, though they looked cheerful.

The Chinese official checked and rechecked the students' documents. Then, struggling with baggage, they were marched down the wharf to the *Star Ferry*. Where's the train to China? thought Anton. For a minute he panicked, wondering how to get Hong Kong money for the ferry fare, but the China official paid. Hurriedly barking commands, he marshalled the group through the ferry turnstile just minutes before its horn blasted and it headed to Tsim Sha Tsui, a few hundred meters across on the northern side of the harbor. Anton was astonished to see an identical ferry pass in the opposite direction. The guide, seeing his surprise, told him the Star Ferry crossed the harbor every few minutes. Then what, Anton wondered, was the hurry to get them on this ferry? It barely docked and they were hustled off. Every thing in Hong Kong was hustle hustle, bustle bustle, rush and hurry.

"Get along there! Get along there!" the official bellowed. "We've got to walk to the train, and it won't wait!"

Walk? Again? To the train? What train? We can't see it! How can we carry our stuff? We just managed to get it on that ferry! Frustration and anger sounded in every student's question.

The guide scrutinized the drooping, complaining group, and decided to hire a small bus. One person would ride with the luggage, he declared, but the rest must walk. Anton resolved to push his bike balancing the precious box containing his German violin and wood working tools on the seat. The other aluminum box could go on the bus. A small red and cream bus was hailed, and luggage piled into it.

"Only large pieces! Only large pieces!" the guide shouted angrily as some students piled all their luggage into the vehicle. There was a scramble to retrieve possessions, and an angry exchange between bus driver and guide. A pale young woman drooping over a large suitcase was chosen to ride.

Finally, the group set off for the unseen station. Despite resting on the seat of his bike, Anton's box leaned heavily against him. Just when panic overwhelmed him, the Hung Hom Station of the Kowloon-Canton Railway came into view. Relieved cries of "Here it is! Here it is!" erupted as the students dropped their baggage on the station platform while their guide bought tickets. The luggage bus was waiting, and the girl did not look quite so pale. The guide returned waving a fistful of tickets, and immediately hustled them on the train.

"Wow, this is a nice train!" exclaimed JingAn, settling into a window seat.

"Sure is!" agreed a chorus of voices.

Luggage was pushed behind seats and up on racks. Anton found space for his bike in the corridor between carriages. The luxury of the train delighted everyone, and they settled down happily for the ride to Canton. There was much laughter and chatter as they rode through the New Territories countryside. Anton knew Hong Kong Island and the Tsim Sha Tsui peninsula were permanently won for Britain by war, but only a ninety-nine-year lease by treaty had ceded the New Territories through which they were travelling. As the train chugged through the attractive green countryside, Anton wondered what would happen when the lease was up. He would most likely still be alive when it expired in 1997, and would learn the outcome. "I wonder where I'll be, forty-two years from now," he thought idly, and grinned at the absurd idea of being so old.

The students were leaning into their comfortable seats, enjoying the ride, when the train halted at another station. "Quick! Quick! Get your stuff and get off! Quick! Quick!" yelled their guide.

They looked at him blankly. How could they possibly be at Canton? Their guide joined the exodus off the train without waiting for them, blinking as he stepped into the burning bright sunshine on the platform. Wildly he gesticulated that they must get off the train. Now! Now! Now!

Amid cries for explanations he angrily shouted, "Get off! Get off! Get off!" Reluctantly they dragged their luggage down from overhead racks, from behind seats, and out of corridors. Barely were they off the train when its whistle blew and it proceeded back to Kowloon. The sun burned down mercilessly. This journey was no longer fun.

"Now, walk over that bridge!" the official ordered, pointing to an extension of the railway track over a very small river. Their surprise could not have been greater. What? Walk again? The bridge was short, but for those expecting a ride to Canton any walk was a shock.

Apprehensively they trudged across the bridge, struggling with their luggage. With relieved shock they rounded a corner and saw another train, a beautiful classic steam engine, although in contrast to the sleek diesel of the Kowloon-Canton Railway it looked archaic. But there was no time to contemplate different engines: their bad-tempered leader was again gesticulating wildly while talking animatedly with a khaki-clad man with a bright red star on his cap. Presently mister star-cap waved them towards the parked train.

"Get on! Get on! Get on!" shouted the guide urgently, and hot, tired, and bewildered, they did.

"What's going on here?" mumbled one of the young women, looking disdainfully at a seat with a large hole in it. "This train is old and disgusting!" The seats had certainly seen years of hard work; many were not only worn but broken.

"It might be dilapidated, but it's a train," muttered a young man, sinking on to the nearest shabby seat. "Better than walking to Canton!" Others muttered agreement, and quickly finding space for their luggage, settled on the battered seats.

The official finally relaxed.

"That was the border," he announced, waving towards the bridge and smiling broadly for the first time since meeting the group. "We can proceed."

The students murmured angrily.

Then suddenly, they understood.

"You are now in China," their guide declared proudly. "China welcomes you. In three or four hours, you will be in Guangzhou!" He looked around the group expectantly. One young man had the sense to clap, and others followed timidly. The guide beamed approval. Suddenly he reached into his bag, and passed packs of cigarettes to each of the astonished students, even the girls. Such a gesture of pure friendliness was totally unexpected.

Anton noticed the uncertain timing of the journey, but after a month of travelling it felt wonderful to be almost at his destination. He placed one of the smooth white cigarettes between his lips, and accepted the guide's offer of a light. He occasionally smoked one of his father's brown *kretek* cigarettes, but the smoke of this Chinese-made version was very strong, making him cough badly and feel very dizzy. But it was a gift, a welcoming one, and he was grateful. The girls put their cigarette packs in their bags, and wondered if all Chinese women smoked.

China was certainly less hustled and bustled than Hong Kong. It was more than an hour before a whistle blew. A few people climbed aboard carrying heavy loads on bamboo poles, clad in somber blue trousers and crumpled white shirts. They sank exhausted into seats, and promptly went to sleep. Finally, the whistle blew, the old train jolted, and with a rattle and sway began its journey north. But the Indonesian students could not rest: their eyes were glued to the windows, absorbing everything they could about this new country. They were so busy looking they had no time to talk, and the journey was strangely silent. Paddy fields turning yellow before harvest and men ploughing with water buffalo were familiar and comforting sights.

Late that afternoon the train slid into the high vaulted grandeur of Guangzhou station. The building was huge but almost empty, as though it had once accommodated much more traffic. Not even the official collecting all their passports and disappearing into the crowds outside dampened the students' happiness at finally arriving in China after their month of travelling.

Another official, clad in ubiquitous Chinese blue, directed them to two small buses that took them to the Overseas Chinese Student Tutoring School in the Shepai district of the city. When they arrived at the school Anton was delighted to find JingAn was assigned the same dormitory as he.

Chinese love ceremony, and cadres had prepared a rousing welcome for these Indonesian students who had accepted the proffered hospitality of the People's Republic of China. Long speeches in Cantonese, of which they understood not a word, then in Mandarin, of which they understood less than twenty per cent, went on interminably. Finally, they were invited to a lavish buffet table, and rewarded with ample to eat. Since they had eaten very little since early morning, most were starving hungry. But the oily Cantonese-style food seemed unappetizing because Indonesian cuisine includes many fresh salads with peanut sauce. To Anton's dismay he discovered many dishes contained seafood. His family ate local Indonesian-style food, but occasionally celebrated special events at a Chinese restaurant where he had discovered eating seafood guaranteed abdominal discomfort and worse. He determined never to eat seafood, and certainly would not risk getting sick at this important time. Fortunately, there was plenty of plain rice.

"KiemGiok," said JingAn as he collapsed on his bunk after the long and eventful day, "this is going to be all right."

"Yes," said Anton, looking for space to store his boxes. "Yes, it will be."

But, he thought, after this long day with so many unfamiliar events, it is very good to have beside me someone who knows I am KiemGiok.

"You know," he said giving JingAn a friendly dig in the ribs, "one name I will never use is my Mandarin name, JinYu, the name my Semarang

Chinese language teacher so diligently taught me! Gold Jade just doesn't sound right in new China!"

JingAn laughed. "A bit imperialistic, eh? You're Anton now, aren't you?" He grinned. "I hope I don't have too much gold in my name!"

*****

The exhausted students quickly fell asleep, but rose early next morning, keen to explore. Signboards announced the unfinished neighborhood was for overseas Chinese who were encouraged to return and rebuild their fatherland. Bamboo scaffolding enveloped buildings under construction. Disordered piles of construction material lay scattered around, dissected by unpaved tracks that led through mud and debris.

Clinging to the edge of this chaos was the Overseas Chinese Student Tutoring School where they would spend the next month preparing for university entrance exams. The school was a motley collection of low brick buildings only recently requisitioned for education, and all exuding the impression that as soon as the students' course finished the whole hodgepodge would be bulldozed into the mud of the construction site. Large notices pinned to doors announced the entrance exam had two categories, Science and Technology, and Cultural Sociology. Anton had no difficulty choosing the science option.

Suddenly, a hand bell rang, and yesterday's guide came running toward them, waving the bell vigorously.

"Come! Come! Quickly, quickly! You must go to the dining hall! Now! Now!" he shouted. "Your devotion to your studies is commendable," he bellowed, puffing from his run and cigarette, the bell tinkling in his hand. "But you must go to the canteen! Now! Right now! Don't you remember I told you last night? You must listen!"

Nodding apologetically, the students hurried to the canteen.

But what were they offered to eat? Last night's feast was not appetizing, but now small chipped porcelain cups surrounded a large pot of dubious white liquid. A large pan contained a glutinous, grey-white substance where floated unidentifiable black and greenish lumps. With relief they saw an enormous wok filled with fried rice.

"Soya milk," the guide announced, pointing to the white fluid. "Very nutritious. Drink it well."

Jabbing a finger towards the glutinous grey substance he declared, "Rice porridge, congee! Today as welcoming treat with one-hundred-year old egg, marvelously preserved by ancient Chinese method of wisdom." Confronted by looks of utter disbelief he added, "It is most delicious."

The students hurried towards the fried rice. "Fried rice!" he proclaimed redundantly. "Mixed with spring onion, egg and pork. Excellent!"

Gingerly, Anton tried the soya milk: it was bland but drinkable. The rice porridge was very salty and the black globs of once-was-egg unpalatable; he was relieved he had taken a very small portion. But the fried rice was delicious, and all the students piled their bowls with it. The official urged them to eat the porridge, but his efforts were futile.

They were given half an hour after breakfast for "personal attentions" and then gathered into a classroom, where they struggled to understand rapid Mandarin and a long list of instructions. Classes were to assist preparation for the "united student examination of all China". Anton was confident he could cope with the science and math examinations, and decided to concentrate on tutorials for Chinese composition, Chinese history, geography and social studies.

Sunday, their day off, the school provided a small bus and a guide to help them explore Guangzhou. The city was crowded with people uniformly clad in dark blue trousers and white shirts. The Indonesians felt very conspicuous. None of them could understand Cantonese, so out came pencils and paper as they had paper conversations with shopkeepers. They were grateful the guide led them to Guangzhou's famous places, Anton's favorite being the Sun Yat Sen Memorial Hall.

Between classes there was a great deal of discussion about which university to apply for. Anton's choice clearly showed his dedication to personal goals rather than prestige.

"You'll be applying for Beida, for sure," his friends assumed. Beida, located in Beijing, was reputed the best university. Competition from students all over China to enter this famous institution was high.

Anton had discovered, however, that for science and technology the new North East China People University (re-named Jilin University in 1958) had an excellent reputation. "No, I think I'll apply for North East China People University."

"What! Are you crazy? Never heard of it!"

"It's in Changchun city. That was the capital of the Japanese puppet kingdom of Manchukuo, where China's last emperor, Pu Yi, had his palace" retorted Anton.

"So?" His friends shook their heads.

Anton wanted science, not prestige. His Semarang school grades showed his science and math were strong, but he had no way of knowing how his grades would compare with Mandarin speaking students. His

Chinese language and history were weak. He would make the North East China People University his first choice, and Beida second. He was convinced it was better to choose what he could achieve, than be sent to a third rate university. "Why fight for number one place?" he asked. "We can't all be at the best place."

Examinations came. A week later results were posted. Fearfully students gathered around the notice board. Some were deeply distraught to learn they had failed. Many were sent to places they had either never heard of or did not want to go.

But Anton was delighted. He had achieved distinction in all his science and math papers, and was assigned to the North East China People University physics department. JingAn was assigned the same university to study in the department of education.

"We've done it!" cried JingAn, slapping Anton on the back.

"We have! We have!" Anton laughed. "And I'm so glad you'll be with me! Changchun is far, far away, you know!"

"Yeah, it is. Even further than Beijing! When do you think we should go? Classes don't start for three weeks."

"Why not go immediately? We can get settled and be ready for classes."

Startled, JingAn looked at his friend. He was thinking of a holiday, exploring China and taking it easy somewhere. Anton was always so focused on his goals! Yet the future was unknown, so very unknown. Maybe Anton was right: find out about their university as soon as possible, and holiday later.

"Good idea," he agreed, looking bleakly at the mud and confusion of the surrounding construction site.

That afternoon they biked into Guangzhou, and bought tickets for the express train leaving for Changchun the following day.

*****

As valued overseas Chinese, although they did not realize this, Anton and JingAn were provided with spacious, even luxurious, accommodation on the train. They had both a seat and a bed, with a clean toilet at the end of the carriage. There was ample space to stow their baggage.

They boarded late in the late afternoon, Anton carefully loading his bike, precious violin, and toolboxes. Soon the sun sank behind the hills and scenery vanished into blackness. They ate noodles brought with them, and looked out into the shades of night.

"I wonder what it's going to be like up there," said JingAn.

"Yeah," said Anton, and fell silent.

What did lie ahead? Changchun was cold, he had read. What was cold like? He had never experienced cold, unless cold was what you called that tingling feeling when breezes blew over the deck of the *Tjiwangi*. And how would he communicate? He could read Chinese, but conversation was very difficult. He glanced at JingAn lying on the top bunk staring at the ceiling of the rocking carriage.

Anton stretched on his lower bunk, and took a deep breath. How could he answer all these questions before he got to Changchun? Exhausted from weeks of intense study, uncertain about his future, and lulled by gentle rocking of the train, he fell asleep. Only a bright sun streaming through the white-curtained window woke him next morning. The train was slowing.

"Changsha!" declared JingAn triumphantly, peering through the window at a station. "Chairman Mao's home! Well, actually, he was born in the village of Shaoshan. Wow, this is exciting!"

Anton sat up straight and ran his fingers through his ruffled hair, perhaps expecting Chairman Mao to walk on to the platform! He had not planned to spend his journey sleeping, and felt decidedly guilty.

"Right," he said. "Let me take a look."

They peered out on to a large station platform. "You know, we're right in the middle of China now. We really are here!" Anton laughed contentedly watching people get on and off the train, many buying food from vendors on the platform.

"I'm hungry!" declared JingAn. He pulled down his shoulder bag and extracted a packet of peanuts.

They had no idea how long the train would remain at the station, and dared not get off. Anton went in search of hot tea. He handed a jar of hot water to JingAn, and got peanuts from his bag. The train started with a jolt, and gathered speed.

"Those trees seem to be running beside the train," JingAn laughed.

"Right!" Anton grinned, looking out. In years to come Anton's chief memory of the journey was trees running by the track. They raced past farmers in the fields, past houses with grey walls and grey-tiled upturned roofs, all so different from Indonesia, so fresh and new to their eyes. But the running trees was the impression embedded in their memories.

Both were deep in thought about what lay ahead of them. But how do you talk about the unknown? An anxious silence wrapped them. Before long the movement of the train lulled their tired minds, and they dozed off again.

Early in the afternoon the train jerked to a halt at Zhengzhou station, waking the travelers.

"So this is what cold is like!" declared Anton shivering, jumping around and rubbing his arms.

"There's no sun outside," said JingAn ruefully. "Grey sky, but no rain."

Anton lifted his small bag of clothes from the wrack and extracted the woolen shirt lovingly made by his mother. He never expected to use it so soon.

"Hey, man! Do you realize we missed the crossing of the Yangtze!" he cried in sudden horror. "We're just sleeping away this journey! Why did you let me sleep?"

"Why did you let me sleep?" retorted JingAn. "Not much we can do about it now! I'm so tired after all that study I simply can't keep awake."

"What are they doing that for?" asked Anton as people pressed their faces against their window and knocked on it.

The boys peered out. The knockers pointed to food in a basket, gesturing vigorously that they wanted to sell.

"They're selling food!" said Anton, in sudden surprise and understanding.

"Come on! Let's give it a go!" suggested JingAn. "I'm starving!"

Anton peered at the hawkers uncertainly. Suddenly his hesitancy vanished.

"They're selling chicken!" he cried. "Just plain chicken! None of that oily stuff we've been eating!"

"Come on! Let's get it!" cried JingAn.

Yanking hard on the frame, they pulled the window up and passed a few *yuan* to the upstretched hands. Neither of them had any idea of prices. Amazingly, a whole chicken was handed to each of them. Ravenously hungry, they leaned back into their seats and devoured it all.

"That's the best meal I've ever eaten!" announced Anton, smiling contentedly.

"I was so hungry I could have eaten that congee stuff they gave us in Guangzhou!" proclaimed JingAn, shuddering in memory of the hated food.

"That chicken wasn't oily like the food in Guangzhou. It was just like the barbeque chicken we ate in Kudus!" exclaimed Anton appreciatively. "I'll never forget this meal!"

"Let's keep awake and see what the country is like," advised JingAn.

But the train's rocking mesmerized them again, and before long both were once more dozing. Several times they woke briefly to nibble peanuts and watch trees run by their windows. When night came they slept soundly.

Next morning, they looked out on a dreary world. A grey-brown countryside, dry and barren, stretched out to a leaden grey sky. A few grey sheep nibbled the grey and stunted vegetation.

"Everything here is grey," sighed JingAn forlornly, and shivered.

The train slowed, and a forlorn grey city marched solemnly past their window. "Tianjin!" declared Anton triumphantly reading the sign as the train slid into a station. "We're getting there!"

A friendly train attendant walked past their cabin.

"Aren't you getting off?" he asked, knowing the boys had not left the train since Guangzhou. They stared at him blankly. Pulling out pencil and paper, he wrote the train would stop for about twenty minutes in Tianjin. As the boys read the note they smiled delightedly, nodding vigorously. The attendant smiled, pointed outside, and continued his rounds.

The boys stood up and stretched. How stiff they suddenly felt! With the guard's assurance they ventured out to buy food. The platform was crowded with a bewildering array of hawkers shouting and thronging around. Much of the proffered food was unknown to them. Eventually they chose pork-filled *mantao* (buns) and loved every bite. The *mantao* were so good they bought more for their journey. Anton was delighted to discover that northern Chinese food was so appetizing. He congratulated himself on his choice of university; at least the food would be good!

As they neared their destination excitement mounted, and sleep no longer overcame them. Silently they watched the strange grey world slide past, each deep in thought about the future. Anton knew it would be at least four years before he could obtain a degree and contemplate leaving China. But he would study hard and make BingGian (and his parents) proud of him!

Shortly before midday they arrived at the city of Shenyang, and late in the afternoon the train finally entered Changchun. Eyes wide with excitement they watched as their new home flashed past their windows. But their interest in scenery had gone: they wanted only to reach their destination. Eagerly they began piling luggage ready to disembark. When the train slid to a halt they were at the door.

JingAn slapped Anton on the back. "We can't sleep in the same dorm here, but let's meet often," he said. "I'd like that."

Anton felt a sudden rush of dangerous emotion. "Sure," he said simply, afraid he might embarrass himself if he said more.

Gathering their possessions, and stiff from much sitting, they climbed off the train, and walked hesitantly out of the station. After the grey countryside, they were surprised the sun shone brightly in a brilliantly clear blue sky. A bitterly cold wind whistled through the station and straight through their frozen bodies. Anton wondered how he could manage to live in such cold. If it were as cold as this in summer, what would it be like in winter? He had spent his entire life in the tropics, and never in his wildest nightmares had he expected it could be so cold. How would he cope with things called ice and snow? The university brochures said winter lasted six months in Changchun. Well, at least there was no ice on the roads, not yet anyway.

Dragging their bulky luggage they rounded a corner, and anxiety vanished: flags identifying different university faculties fluttered at the station entrance. Anton and JingAn had go their different ways, but the university representatives were warm and friendly. It was a good start.

Anton was directed to a small bus that took him to the hostel of the physics department. Seated on the front seat beside the driver he was astonished by the city's broad main road, Stalin Street, so broad it was divided into three parts by two rows of magnificent, lofty trees. The central lane was for motorized vehicles, the two outer for bicycles, animal-drawn carts, and pedestrians. He noted with surprise there were no overhead electric wires, and was amazed to learn these were buried underground. Although heavy industry was the foundation of the city (it boasted the first motor car and tractor factories built by the Chinese government under Russian direction) the driver declared it was a clean and safe place.

Anton was glad he still had a couple of *mantao* from Tianjin, and could concentrate on getting his things in order instead of going to the canteen. He did not feel ready to face a new crowd of people.

Next day he was given a tour of the facility and allocated textbooks and other essential needs. He was most grateful for the thick padded trousers, coat, and hooded "monkey" jacket that were distributed to all new students. It solved his immediate worry about coping with the biting cold (although later he discovered nothing kept out the aching Changchun winter cold).

But most unexpected, proving a real struggle, was food. Anton's acceptance into the university meant he was supplied with accommodation (a bunk in a six-bed dormitory), books, a yearly issue of clothing, a small allowance of 14.50 *yuan* a month to spend as he liked, and food at the faculty canteen. In Guangzhou there was always plenty of rice, and, apart from checking for seafood, eating was no real problem. Anton enjoyed the chicken, *mantao* and noodles of northern China. But in northeast China the staples were maize corn and red sorghum. Anton struggled for years to even eat these grains let alone enjoy them; in fact, he never did. At least he had no worries about seafood in Changchun!

Thus, just after his twentieth birthday, in late August 1955, Anton began the four-year basic physics degree of North East China People University. It was comprehensive and thorough. In the first year it included chemistry, math and electronics. The next year covered mechanics, thermodynamics, and quantum physics. The last two years focused on higher math, such as Rieman geometry, Einstein's theory of relativity, more quantum theory and nuclear physics. Anton discovered the teaching was based on the Russian system, and all textbooks were translated Russian texts. The classrooms were bare and basic. The laboratories had minimal equipment (not even as much as his Semarang high school), but the students faithfully learned everything in the Russian texts. Many learned text and formulae by rote, but just as he had done at middle school in Semarang, Anton always looked for the principles behind formulae.

The first year was tough, tougher than his wildest nightmares. His deficient Mandarin meant he understood little of classes, and joined in few conversations. He studied the texts diligently, but was often puzzled to understand lectures. Eventually, with profound gratitude, he discovered his classmates were willing to give him help. Fortunately, his math and physics were strong enough to cope, and this slowly led to his acceptance by students and teachers. A significant breakthrough came one day during a discussion forum. The Russian university system then used in China consisted of lectures given to large classes by the professor, medium classes taught by a lecturer, and small classes by a tutor. Discussion forums were smaller groups designed to help students solve difficult problems. One day Anton dared to join a discussion forum dealing with forces.

The tutor drew a diagram on the blackboard and asked the class to find the resultant forces. For five minutes' no one made any response, yet Anton was sure he could solve the problem. It was similar to exercises he had done many times in the Semarang Chinese English School. He raised his hand, went to the blackboard, and silently and quickly solved the problem, writing his solution carefully in English. He turned to face the class. A sea of silent, blank faces stared back at him. Eyes bored into him as he slunk to his seat, certain he had made a fool of himself. Suddenly the tutor, very surprised by Anton's use of English, quickly praised his effort.

"You see, Ang Tong has shown the way!" the lecturer said, in simple Mandarin even Anton could understand.

After this, both lecturers and students took more interest in this "dumb" student from overseas. As they began talking with him his facility with Mandarin increased slowly but steadily. They told him most overseas Chinese students did not have a good standard of learning, and they assumed he was silent because he knew nothing.

*****

To his surprise, Anton found compulsory political classes interesting. Many of his classmates thought of themselves as scientists and were bored with political discussion, but Anton welcomed the chance to learn things he had never thought about before.

The course was divided into two parts. First were political seminars, in which Anton learned the details of recent Chinese history, Russian history, revolutions, and the ideology of the Russian communist party. Russia was then China's only significant ally and was both political advisor and financial supporter. Karl Marx's *Communist Manifesto* was a major focus of study. To emphasize the legitimacy of the present Chinese government, the history of a wide range of world revolutions were studied.

The second part of political study was primarily philosophy. This made Anton reflective about his life and values, and developed his strong interest in philosophical and religious ideas. The course covered the basic ideas of Objectivism and Subjectivism, the concepts of Dutch philosopher Baruch Spinoza and German philosophers Georg Wilhelm Friedrich Hegel, Ernst Mach and Friedrich Engels. Chinese philosophy centered on the essentials of *Yiching* (the Book of Changes), and the ancient writing *Sun Tzi*, known in the West as *The Art of War*.

Anton thought these classes were a pleasant change from his rigorous mathematical and scientific course. Years later, he believed this training in philosophy enabled him to solve problems in quantum mechanics and relativity, and helped him to contribute to the understanding of violin acoustics. Since he was a man of few words, he rarely expressed his personal views about political and philosophic ideas, but he was glad to have the chance to learn more and to think objectively about his work.

After graduation the political study groups continued. Anton was interested to hear the ideas of others, but when the political situation during the Cultural Revolution became volatile, his habit of quiet reflection was an effective safeguard. He developed his own philosophy, and his own standards. He decided that both political and personal actions were either good or bad. He defined "bad" as selfishness, laziness, greed and corruption, and "good" as working hard, thinking more of others than self, and giving respect. He made it a cardinal rule never to force others to do something. He developed a simple "formula" for right living: give 110 per cent to others, and keep 90 per cent for himself. Peace, developing respectful and harmonious relationships with others, were what he valued. He was later shocked to discover that few people with the so-called freedom of western thinking understood the importance of unselfishness and harmonious relationship.

Politics, however, he decided was a game, one that he could not play because he had no skills for it. He preferred to make friends of all nations, people of all religions. He later believed this was the primary reason he had no trouble during China's difficult years.

*****

With growing confidence, Anton ventured to get involved in other university activities. Because he had enjoyed ping pong in Semarang, he joined the table tennis club: it was the only sport he knew anything about. Soon he enjoyed being recognized as an expert in this game.

A few months later in the winter term he surprised himself by winning the university inter-faculty table tennis championship. A month later he became the champion of Changchun city. Now many people noticed this silent, serious young man.

"Where did you learn such good technique?" his classmates asked. "Can you teach us to play ping pong?"

"Of course, but I only play for fun. I'm not trying to win or be the best. Trust me! Why don't you hold the bat this way? Gentle hits are sometimes more effective than strong ones." He enjoyed coaching his friends.

He joined the Performing Arts team, and played either violin or guitar to accompany dances. The university encouraged ballroom dancing every weekend as a way of connecting students, and these dances were very popular. They played Strauss waltzes, Hungarian and Romanian music, and syncopated Latin *tango* and *milonga* music. Strictly forbidden were jazz and Hollywood music. Other cultural activities offered were Chinese opera and ballet.

Anton's social life in Changchun slowly blossomed, but its impetus was tragedy.

Towards the end of his first year he received one of BingGian's frequent and welcome letters. But this time it deeply shocked him. She began by saying her studies were going well, no surprise. Then he was startled to read her suggestion that he should find another girlfriend because she was not good enough for him. Just as he was mentally composing an ardent eulogy of his appreciation of her he was devastated to read that BingGian was already married. BingGian married? He could not believe it. Married to someone else? Impossible! Why, he studied hard for *her*.

He read and reread the letter. It was her handwriting, of that there was no doubt. It must be true. And there was nothing, absolutely nothing, he could do. He was plunged deep into misery.

But after a few days he realized he did not want someone who did not truly love him. He should get on with his studies. He knew she had inspired him to do his best in Semarang. For this he was forever grateful. Finally, after several weeks, he wrote BingGian a gentle letter wishing her all the best.

Yet the aching hole in his heart did not disappear. He threw himself into his work, and helped his classmates as usual. The government encouraged gender equality and there were four girls in his physics class. One who frequently needed his assistance was Tang ZhiXiu. She seemed kind and attractive, and their relationship deepened. One Sunday she told him she had received a letter from an old school mate who was coming to Changchun and wanted to visit her. She asked Anton's permission to meet him.

Anton was a little surprised at the request, but responded cheerfully, "Sure, go and see him." He was busy with some experiments that were not working out well, and so saw no reason to accompany her.

But after the visit he sensed an increasing barrier in their relationship. Often he felt she wanted to talk, but was unable to express herself. Finally, a month later, he was very surprised to receive a letter from Tang ZhiXiu's old classmate. He declared ZhiXiu was his girlfriend, and had been for many years. This time Anton was not sad but angry.

As soon as he could arrange to meet ZhiXiu, Anton confronted her with the letter. "Read this!" he commanded. "Is this true?" he added, coldly.

Miss Tang hung her head and did not reply.

"You need to make a decision," said Anton, his voice sharp with pain and anger. "You can't have both of us. It's either him or me."

ZhiXiu wept and said she cared about them both, that she didn't mean to cause trouble, that it was all a big mistake, that she really did care for Anton, and numerous other protestations. Anton sat silently and waited.

"But what is your decision?" he quietly demanded again.

Miss Tang knew that Anton was kind, gentle and extremely patient. Although he was the best student in the class he never clamored for attention, never tried to force his ideas on anyone. He had been a big help to her when she was struggling with her studies. But as she sat awkwardly twisting her hands while he waited for her decision she became aware of his eyes, eyes that both his daughters later commented on, eyes that seemed to bore right through her. She knew that he demanded of others what he demanded of himself: the truth, the whole truth, and nothing but the truth; commitment, total commitment, and nothing less. She must tell him the truth. She had treated him very, very badly, had completely betrayed him, and she knew it.

Finally, she stood, turned to the door, and blurted out defiantly, "I don't think you will ever trust me, so I choose him." She walked out, slammed the door, and fled.

For a week Anton was distraught. Life was cruel. He was angry with both himself and Tang ZhiXiu. How could he have been so blind? Finally, he decided it was good he had found out the truth. If she could not be true to him, they could never have a meaningful relationship. Perhaps he should have gone with her to meet the old "classmate." But a forced relationship would never work. It was best ended so he could get on with his life. But despite his efforts at rational thinking he was deeply hurt. Again he poured everything he had into his studies. He could trust his work, but he could not trust relationships.

*****

Just when he felt he could meet ZhiXiu in classes without anger and pain, Anton received another shock. In May, towards the end of his second year, he collected a very official letter, smothered in large red chops, from the university office. Apprehensively, he tore open the impressive envelope. His roommates crowded round to discover why he should get such an unexpected document.

"It is with great pleasure that the Comrades of the Science faculty inform you that you have been selected to leave your studies and join the Jilin Province Table Tennis Team," he read.

Anton stopped and stared in shocked horror at the page. He read the words again, to make sure he was not having a bad dream. "This can't be true! Someone has made a very bad mistake," he cried. "It simply can't be true!"

"What's happened?" asked his friends, crowding closer and hitting heads as they tried to read. "What's wrong? What's happening?"

"They must be joking!" Anton gave a hollow laugh. "They can't be serious! Who said I want to be a sportsman! They've made a mistake with the name!"

"Not likely!" his friends laughed. "No one has a name like you." The young men read the letter spread out on Anton's desk.

"Wow!" they said in chorus. "Lucky you!"

"What an amazing opportunity!" added one. "You'll get to travel all over China, and be paid for having fun! How lucky can you get! Wish I'd taken ping pong lessons from you!"

"But I want to be a scientist! I've no interest in sport! It's only fun! Sport isn't real. They didn't even ask me!"

His roommates looked at him in amazement. "How can you be so crazy to even think of refusing such a chance?"

Anton turned back to the letter. It explained he would receive intensive training, and would take part in national sports competitions. There could even be the chance of international contests if he proved good enough.

Anton took a deep breath. The directive was supposed to be an honor, but for him it was a crushing burden, the shattering of all his dreams. He had no interest in professional sport. He had not even the slightest wish to be a fulltime table tennis player, or perhaps, even worse, if he was not good enough to perform, to become a professional referee. The letter was the death knell to everything.

"Hey, can any of you play ping pong?" he asked hopefully. "Maybe we could do a swap?"

But, in the China of 1957, a directive from the authorities was an absolute command. There was no appeal. The university was so convinced that this was a tremendous honor for both Anton and the school of science that they were unwilling to petition on his behalf.

Life now seemed tough, strange, and utterly beyond one's control. One girlfriend had married someone else, his latest girlfriend had double-crossed him, and now he was forced to leave university just because he happened to be good at table tennis! With a heavy heart Anton obeyed, reluctantly reported for ping pong duty, and embarked on his career in table tennis.

But, he thought, they might take science away from me, but they can't take music! As he packed his tools and violin into the trusty aluminum boxes he thought, "Well, I won't ever be an Einstein, but maybe somehow I can still be a Stradivarius!"

Within a few weeks, however, the trainer noticed that his champion player, winner of the North East China People University table tennis championships, and Changchun city-wide competitions, was coughing badly and not performing very well. On questioning Anton admitted he had been unwell since winter, as always intensely cold. Temperatures plunge as low as minus 35 degrees Celsius in Changchun. Although all housing was equipped with wood or coal-burning stoves, or in the more modern larger apartments piped steam heating, it was bitterly cold outside. No one could avoid the cold walking to classes or the canteen. Thickly padded clothing helped, but the freezing air still had to be breathed!

A medical examination quickly revealed Anton had pleurisy and a pleural effusion. He was admitted to hospital, and treated there for three long, tedious months. He used his sudden free time to study science and music theory. The nurses allowed him to practice his violin in the ward office when they did rounds with doctors. Fellow patients were amazed at his dedication. Although Anton was never told his diagnosis, tuberculosis was the most likely cause of his breakdown in health.

Eventually a university official in a crisp new Mao suit came to the hospital. After several minutes of awkward small talk, he informed Anton that his services to table tennis were no longer needed. "I am very, very sorry to pass on this bad news," he said, handing Anton another letter loaded with bright red official chops.

He was astonished to see Anton's face suddenly beam with joy as he read the letter. "I can go back to university, sir?" Anton asked cautiously.

"Yes, the privilege of representing your county can only go to the best. You are no longer needed for the table tennis team."

"Yes, sir," Anton replied diffidently.

Anton managed to sit quietly until the man left the ward. Then, grabbing his violin, he swung it in wild hoops around his head and danced riotously. This wild performance from the usually quiet Anton stunned the other patients. Suddenly they all began clapping. He stopped, bewildered, and looked at them. Their clapping died away, but their faces were kind and friendly. Anton tucked his violin under his chin, retrieved his bow from the bedside cabinet, and began to play the magical piece he had heard Fritz Kreisler play those long years ago in Kudus. As the last notes of Dvorak's *Humeresque* died away, the ward erupted in joyful shouts and more clapping. The patients had no idea why he was so happy, but they happy were for him!

Anton had missed a whole year of study, but his illness had saved him from the dreaded fate of professional sportsman. It is also likely that the table tennis and hospital interlude saved his life. As a mere student his health issues may not have been observed as quickly, or treatment initiated as effectively. And perhaps, it was the extended hospital spare time that allowed his intensive study of music theory and harmony that later proved vital for him.

*****

In the autumn of 1958 Anton returned to university. The past year had been devastating, but now he could gratefully start again. Before his excursion into table tennis champion he learned he had passed the year with distinction. If he could keep this up it would assure his place in the two-year advanced program after graduation. For the advanced program he would specialize in metal, magnetic, semiconductor, and theoretical physics. Although things had gone wrong with his relationships, and he had missed a whole year of study, he could once again immerse himself in physics and music.

But China, which had seemed to him a wonderful Shangri-La of care, security and safety, was changing. Strange political winds were blowing. Everywhere were slogans declaring that China would overtake the economic status first of England and then the United States. Amazingly, as part of this plan, communes were said to be making steel right in their back yards! But along with these amazing feats were reports that intellectuals were frequently denounced. Some teachers at the university just disappeared. Fortunately, no one from the physics department left. Like others in the university, Anton decided the best thing to do was put his head down, say nothing, and get on with learning.

The most difficult immediate problem at this unstable time was food. Students were completely dependent on the university canteens, and the quality and quantity of food deteriorated daily. Anton even began to long for the dreaded red sorghum. They were supplied with "supplementary" food items: dried poplar leaf, dried corn leaf, cassava root mixed with sorghum, and small allocations of maize corn and rice. Everyone was given a ration of food stamps, so there was no chance of eating more than one's allowance. Hunger was a daily constant. Vegetables and meat almost disappeared from their diet.

One day, Anton, after completing some tricky experiments in magnetics, got ready to leave the laboratory. The textbooks gave an excellent grounding in theory, but the labs had only very basic equipment. Achieving good results from experiments was always a challenge. He was surprised to collide with Professor Sun ChiaChung as he walked through the door.

"Still working?" asked Professor Sun. "You're very conscientious."

"I try to do my best," answered Anton. "This experiment has been difficult."

Professor Sun nodded. "I understand. I have something for you," he said, suddenly bringing a cloth bag from behind his back. "I think you need this."

He thrust the bag into the astonished Anton's hands. Smiling, he quickly walked away before Anton could say a word.

When Anton opened the sack in the privacy of his room, he was amazed to find inside a whole cabbage. Tears sprang to his eyes. Surely, he thought, this is the most perfect gift I have ever received. He did not dare show his friends for fear of revealing the donor. In the strange political world that China had become, he strongly suspected it was not acceptable to give food. Quietly, whenever suitable in the privacy of his room, he ate a little finely cut raw cabbage. The irritating bleeding patches on his skin disappeared.

It was several years before he began to piece together facts that gave understanding to what happened during the years of hunger. He knew there had been some natural disasters, but why had the hunger gone on and on? Then one day, as he stood with his elder daughter near the busy Changchun railway station watching trains come and go, she asked, pointing to an enormously long goods train, "Where is that big train going Daddy?"

Casually he answered what everyone knew, "To Russia, child."

Then suddenly, he understood. Russia had given a great deal of help, advisory and financial, to China, but the friends had argued and became enemies. Russia demanded that China repay debts that China thought were gifts. The enormously long goods trains were filled with grain for Russia, grain that would repay the debts, but would not feed the hungry people slaving on their communes in China. Anton knew he had received a Russian-style education and was grateful for it, but he had serious doubts about Russian morality. This realization strengthened his belief that it is unwise to accept favors from others. He often admonished his daughters that they should never take gifts and benefits from others. He strongly emphasized the importance of being independent.

In the 1960s Anton received a letter from his parents that dispelled any concerns he had about lack of food, the political scene in China, and his choice to study there. His younger brother, KiemSiang, had also won a scholarship to the Chinese English School in Semarang. He excelled in chemistry and geography, and his teachers were confident he could gain admittance to any university in the world. KiemSiang considered his options, and decided on Bandung University in Indonesia so that, with filial devotion, he could remain near his parents. But his Chinese ethnicity proved a discriminating handicap. He was forced to leave university and go to work in a textile factory. He had no chance of further education.

Anton felt very sad for his brother, but he thought he understood why Indonesians were angry with Chinese people. He believed colonization everywhere followed a similar pattern. It began, he noted, with trade, but soon a middle agency was needed to control the native people and make trade safe and profitable for the ruling country. The middle agency was usually a previously powerless minority group in the colonial country. Under the Dutch rule of Indonesia Chinese were the agents used to control and develop industry and trade. Many Indonesians considered the Dutch were their friends and the Chinese agents the cause of all their troubles. His brother, he believed, was a victim of this situation.

In his last year of university Anton joined the University Student Music Tour as a violinist. The orchestra gave performances in Harbin and Dairen, and had a great deal of fun. Some of the performers noted his expertise on the violin and asked for tuition, which, as usual, he was glad to give. One of them, Huang WenXue, later became a violin teacher in the Shenyang Music Conservatory.

In the summer of 1961, Anton graduated with distinction, the proud possessor of an advanced degree in physics (the equivalent of a doctoral degree). Colleagues called him Dr. Sie.

But what brought him most joy was his appointment to the position of Researcher for the Physical Research Unit of the university. It was his dream come true, the end point of all his plans and aspirations since middle school in Indonesia. He was a recognized physicist.

The position involved undertaking research planned by the head of the physics department, presiding over forums and teaching graduate students. Anton enjoyed his studies in physics.

The university noticed his interest in violin acoustics, and his desire to discover the secrets of violin construction perfected by the Cremona luthiers. Professor Sun agreed he could use his free time to start his own research on the structure of violins, to see if he could learn factors that might lead to differences in the tonal quality of instruments. The city of Changchun was conveniently close to the Changbaishan mountains which were heavily forested with good quality trees. Anton planned to use his days off in summer to visit the mountains and obtain wood suitable for violinmaking.

But to work with wood meant Anton needed equipment. In the grounds of the physics department was a small iron foundry, run by an old master metalworker called Li. Anton developed a good relationship with this skilled old man, who helped him make three top quality hand-tempered chisels. Their quality proved to be so good that fifty years later Anton still uses them. Mr. Li did not have a prestigious position in the physics department, but his skill and attention to detail caused Anton to deeply respect him. He remembered Mr. Li with gratitude and affection for the rest of his life.

Despite working with the Balinese wood carvers in Semarang and learning all he could from them, Anton realized he needed to improve his woodworking skills. Another university employee was a master woodworker called Qi FungWu, responsible for making wooden models of machines designed in the department. Anton befriended Qi, and learned everything he could from this accomplished craftsman.

To further his violin research Anton studied what others had discovered about violin acoustics. The library of Jilin University was large and well stocked. Anton visited it regularly. He became good friends with the librarians due to skills learned in his Chinese English School days: twice a year he helped the library office service their typewriters. In gratitude they allowed him unlimited reading and borrowing of books and magazines. Two magazines he took copious notes from were the German *Zeitschrift fur physic* and that of the American Acoustical Society. He studied the violin making techniques of Edward Heron-Allen and Karel Jalovek. The Heron-Allen book he had brought to China in the aluminum boxes. He also studied the harmonics of great composers, particularly Bach, Handel, Mozart and Mahler. Everything about music that he could find in the library he eagerly devoured.

However, the library tempted him to expand his horizons in other areas. He learned basic German and Russian so that he could read music and physics magazines in these languages. He maintained his English by reading English books and magazines, though he had no opportunity to speak the language. He read great literary masterpieces, for example the works of Dumas, Tolstoy, Shakespeare and Hemingway, naming a few to indicate the variety of his interest. The changing political scene of China encouraged reading in philosophy, and he devoured the thoughts of Descartes, Spinoza, Hegel, Engels, Malthus, Lenin, and, interestingly, Abraham Lincoln.

Philosophy led to religion, despite its being proscribed by the Chinese leadership. Quietly working in this great library of China, Anton read about all the world's major religions. Like most Indonesian Chinese, his family had not been particularly religious, although Chinese culture was important to them. But Anton realized people needed guiding principles in life. Kudus was a religious city, and proud of it, yet the Chinese living there had felt under no pressure from their Muslim friends to conform. After much reading Anton concluded that for him the teachings of Lao Tze, known as Taoism, made most sense, although he also valued the teachings of Jesus Christ. He decided on two guiding rules for his own life: he should not force others to believe as he did, and peace was much greater than fighting.

Anton's general thirst for knowledge extended to things medical, stimulated by his three months' hospital experience. He became acquainted with a teacher of Chinese medicine, Dr. Hu, and under his influence learned the basics of Chinese herbal medicine and acupuncture. He became quite proficient in the use of acupuncture, sometimes using his skill to effectively treat friends and family. Some people called him a barefoot doctor, which both amused and pleased him.

*****

In 1962 Anton married, an event that ultimately changed the course of his life. Peace-loving Anton may have remained a physicist taking directions from his superiors in a provincial Chinese university, but tragedy led his wife to encourage him into a different course.

Wang YuXiang's workroom was close to Anton's university office. Although not university educated, she was a highly skilled technical draughtswoman, much in demand. Her work was beautifully accurate, and her calligraphy drew admiring comments. Anton noticed young men found excuses to spend time in her office. YuXiang, oblivious of their attention, remained aloof from them all. She was tall and attractive, but after his previous disappointments Anton was in no hurry to begin any relationship.

One day he took some work to YuXiang for drafting. A page with some of her calligraphy lay on the bench.

"Calligraphy is like fine art," Anton commented philosophically.

YuXiang looked up. This comment was very different from the banal observations she usually heard about her looks or the weather. "Are you interested in art?" she asked.

"Yes, my father enjoyed art and I won prizes at school for painting and drawing. In Indonesia I worked in a photographer's shop coloring photos."

"Really? I would love to do that." Her interest surprised Anton.

A few weeks later he returned with more plans for copying.

"I am really interested in learning to paint pictures," YuXiang admitted shyly, taking his plans. "Would you be willing to teach me?"

Anton, used to people asking for his help, did not hesitate to agree.

YuXiang was an apt and diligent student. But Anton was busy with his new research position and preparations for his violin research. He was merely happy to be helpful. The "classes" occurred during lunch breaks in her office, and never lasted more than fifteen minutes.

YuXiang was not talkative, but she asked questions about Indonesia, and how Anton came to be in Changchun. She showed a strong interest in learning things beyond the narrow confines of her own life. She discovered they shared a mutual interest in music, although she did not play any instrument. One day he told her how ping pong had interrupted his education and to his surprise she frowned. Other people had always thought he lost a wonderful opportunity.

"I am glad you were able to do what you really wanted to do," she said. "Our country has given me good opportunities, but it is good to do what you really like. I come from a very small village in the country. I did my best to learn, and had the chance to do this job. You are fortunate that you could do what you wanted!"

Anton was intrigued. She seemed just an attractive girl who wrote well, and liked painting. Suddenly he realized she had dreams and hopes. She might not be university educated, but she was interesting.

"You had to leave your country and travel far to get an education, but I am glad."

Before Anton could reply, YuXiang gathered her papers and paints and ran from the room. He was shocked to see she was crying. The following week the painting tuition continued, and YuXiang did not allude to her dreams. She asked simple questions about other countries, but revealed no more about herself.

His surprise therefore was unbounded when one day she handed him a large, heavy parcel.

"It's a gift to thank you for your help in teaching me," she said shyly.

With classic Chinese politeness Anton did not open the gift in front of her, but his astonishment only increased when, in the privacy of his room, he saw she had given him a magnificent book of paintings, *The Best Collection of Chinese Paintings*, clearly a very expensive gift. Years later, he discovered YuXiang had spent almost two months of her salary on the book. This young woman clearly appreciated him. But he did not think she could possibly be seriously interested in him as he had few of the material advantages young women at that time expected in a marriage partner. He had long since sold (to cover expenses beyond his monthly student allowance) the ancient bike brought from Indonesia, and he had not been able to replace it. Although he had an old watch and a radio, he did not have the means to buy a sewing machine, and thus owned only two of the expected basic four requirements for marriage, according to his friends. He put the thought out of his mind.

But he could not resist learning more about YuXiang. Her family was originally from Shandong and had migrated to Changchun early in the century. Her father had died when she, the eldest of three daughters, was very young, and her mother remarried. Her stepfather was a kindly factory worker who regularly received model-worker awards. Her mother quietly developed small enterprises to augment the family income. She kept rabbits, and supplied friends and acquaintances with rabbit fur and skins, very much in demand cold Changchun. She managed to grow *clivia*, exotically beautiful lilies, for which people paid a fortune at Chinese New Year. From her mother YuXiang developed an extensive knowledge of plants. Although both YuXiang's parents were illiterate, they sent their three daughters to school.

Anton admitted Wang YuXiang was a remarkable young woman, and like many others he began to find excuses to visit her office.

Eventually YuXiang invited Anton to her family home in Er Dao He Zi. The home was a traditional northern Chinese house: single storied, with a large knee-high brick platform (the stove) incorporating a huge wok and a small opening for firewood. The stove's low height was necessary because the chimney from the stove passed through the wall to heat the *kang*. A *kang* is the center of a traditional northern Chinese house, and the only essential piece of furniture. The family sits on it, entertains guests on it, eats on it, and eventually, rolling out quilts stored beside it during the day, sleeps on it. There are no chairs and no tables. Anton quickly discovered YuXiang's mother was an extremely good cook, and better, had taught her daughter her skills. The university canteen's tasteless cornmeal bread normally eaten merely to satisfy hunger was suddenly a gourmet food when YuXiang's mother made it.

The more Anton got to know YuXiang the more he appreciated her beauty and kindness. Her fine analytical mind, and general curiosity about life impressed him. She also had the classic virtues of a good Chinese woman -- she was modest, considerate, caring, truthful, and gracious. The gift of the expensive art book indicated her willingness to sacrifice for him, but in many small and simple ways she was kind and helpful to others as well himself. Although cautious after the distress of his two previous relationships, he realized this young woman really loved him, despite his lack of sewing machine and bicycle. Their willingness to marry was mutual.

Anton had continued to live in the university dormitory, normal for all young single men. China had a serious housing shortage, and many couples began married life with both partners still living in dormitories, waiting for their names to slowly rise up the long waiting lists for an apartment. But when Anton approached his *danwei* (work unit, or community) he was delighted to learn that when he married he would be allocated an apartment in housing Block 4600 situated on Jie Fang (Liberation) Avenue, in the Nan Guan district of Changchun. It was an older housing block, but it would be their own personal home. It had hole-in-the-floor communal toilets (men on one side of the central staircase and women on the other) and a large communal kitchen, but both Anton and YuXiang were delighted to have a place to call their own. They married in 1962. In 1963 their first daughter was born, and they named her NeeZi.

Food shortages continued, and coupons controlled all food purchases, but the government provided basic necessities. At 58 RMB (*yuan*) per month Anton's monthly salary was almost luxurious, compared with YuXiang's regular worker salary of 40 RMB. They paid a ridiculously small 60 *feng* (cents) per month for housing. Health care was entirely free, including hospital expenses. YuXiang, after giving birth, was entitled to extra egg coupons, and an allocation of soy bean powder to make milk for NeeZi. They knew they had a low standard of living, but Anton believed it was all they needed, and felt very secure.

*****

In the summer holidays of 1964 Anton received another summons from Beijing, this time one that gave him great joy.

He arrived home jubilant. "Look YuXiang!" he called, opening the door and waving an impressive-red-chop-covered envelope. "Look what I've got!"

"A new job? Money? A change of housing?" YuXiang hazarded guesses wildly.

"I'm to be a translator! I can meet fellow Indonesians and show them China! I never ever expected such a privilege!" He placed the letter front of his wife.

The Central Cultural Committee directed that he proceed immediately to Beijing, to be tripartite translator for the Sukarno Cultural Delegation that was making a tour of China. Anton's proficiency in Indonesian, English and Mandarin Chinese was in demand. Although others had come to China from Indonesia, clearly the authorities had confidence in him.

"Really? How long will you be away?" YuXiang asked anxiously, as she fed NeeZi soft cornmeal mash.

"Oh, not long. This is not a permanent arrangement. But it will be good to hear *gamelan* music again." Anton pointed to the dates he was required for translation duty.

YuXiang looked down at NeeZi. "They make so many demands," she said flatly.

"But this is such a privilege!" said Anton, surprised at her reaction. "I never expected such an opportunity!"

"True," she answered. She was pleased Anton had this opportunity, but wondered how she would cope alone with both her work and NeeZi.

For one month Anton travelled China with the Indonesian group. They gave performances in Beijing, Shanghai, Hangzhou and Guangzhou. It was pure joy to hear traditional *gamelan* music, and to assist the group. When in Beijing he was delighted to meet Chairman Liu Shao Qi. The thought of a closer relationship between China and Indonesia filled him with hope. Both the Indonesian group and the China Cultural Committee expressed appreciation for the assistance he gave. Anton returned home to YuXiang and baby NeeZi full of joy and optimism.

But the following year Anton was horrified to learn that a new Indonesian president did not value improved Sino-Indonesian relationships. Chinese were blamed for political uprisings in Indonesia. Virtually all members (mostly Indonesian) of the touring cultural team were shot. Anton could not begin to understand the motivation of such wanton destruction. His revived hopes for reconnecting with his family were dashed.

Soon after Anton returned from his tour with the Cultural Group he arrived home with more good news.

"Start packing!" he announced to his wife cheerfully, noting her suddenly anxious expression. "We have to move."

"Oh, no," groaned YuXiang.

Grinning broadly, Anton added, "We've been allocated an apartment in the brand new Block 2200 across the road!"

"Oh Anton, have we really?" YuXiang beamed.

Yes, he nodded. Very excited, YuXiang seized NeeZi from her cot, grabbed Anton's hand and swirled them both around.

"Oh Anton! This is good fortune. You must be making someone very happy!" YuXiang was well aware her husband worked in a privileged *danwei*.

Their new apartment, on the second floor, consisted of two warm, light, south-facing rooms. It had a main room for sleeping, eating and working, and a kitchen shared with only the neighboring family. They even had their own toilet-bathroom. Moving was not difficult; they merely carried their few possessions across the road. The apartment had a balcony that operated as natural freezer in winter, and chicken coop in summer. YuXiang could now prepare small amounts of green beans, eggplant, and chili to dry on ropes on the balcony-cum-freezer, which gave variety to their austere winter diet.

Like all Chinese women, YuXiang returned to work soon after the birth of NeeZi, who was left each day in the care of the kindergarten on the ground floor of Block 2200, part of Anton's *danwei*. YuXiang was extremely grateful of this convenient childcare.

Anton went to work in a shuttle bus provided by his physics department *danwei*. But even though she worked at the same place, YuXiang was not so privileged. Her ordinary worker *danwei* did not supply a bus, and was a daily reminder of the educational difference between them. Although public buses and trams were available, she, like almost everyone in Changchun, preferred to walk to save money. It took her thirty minutes each way. Considering the extremely cold winters of Changchun this was a significant hardship. Often she slipped on the icy roads, and every year developed painful chilblains. The twice daily walk demonstrates how much she needed money and how determined she was to conserve their meagre financial resources.

NeeZi was a cheerful child, a joy in the family. Her gleeful singing alerted the whole apartment block of her return home. She mounted the stairs singing lustily the latest political song being broadcast on public radio. In summer climbing the stairs was fun, but it was challenging in winter. During the long winter months stairs doubled as storage space. Hundreds of kilograms of cabbage and potatoes, procured in late autumn to last all winter, were piled on the stairs. Large tanks of pickled vegetables also blocked stairs and corridors, along with chicken coops moved from the icy balconies to take up residence by toilets. The smell of human and chicken excrement plus pickled vegetables was powerful, but all tenants thought it a small price to pay for precious vegetables and eggs. Among all this clutter a few bicycles found space to lean against the walls.

In winter this disorder was negotiated in dim light or even darkness. Whilst good for playing hide and seek, small girls were often terrified. The government provided all maintenance for the apartments, but changing broken light bulbs stretched their capacity; broken bulbs were rarely changed before summer. Tenants, including children returning from kindergarten, had to manage as best they could.

Uncharacteristically, NeeZi arrived from kindergarten one day crying inconsolably. It took several minutes to soothe her.

"My egg!" she wailed between sobs. "I lost my egg!"

"Your egg? What egg?" YuXiang was puzzled.

"In my soup! My egg! They gave it to a boy! A horrible, nasty, greedy boy!"

YuXiang discovered that for lunch that day the kindergarten had provided spinach soup with egg. NeeZi hated spinach but loved eggs. She decided to eat the detested vegetable first, and leave the precious egg till last. But hated spinach just could not be swallowed. Slowly, painfully slowly, she forced each detested mouthful down, but she was too slow. The ill-tempered kindergarten teacher decided she was not hungry. Her bowl was snatched away with little but the precious egg remaining, and given to a boy NeeZi did not like.

"It was my egg! I told her I wanted it!" NeeZi sobbed. "She gave it away ... sob ... she gave it ... sob ... to that horrible boy!"

YuXiang hugged her tightly, but to this day NeeZi remembers the incident with a terrible sense of loss. She came to a firm decision: she did not like boys!

*****

Even Anton, absorbed in his research, noticed his small daughter's singing. Her enthusiastic chanting of political songs was a benefit to the family, especially during the Cultural Revolution. It publically declared to everyone that the Sie family were loyal supporters of the government. YuXiang needed only to get NeeZi to repeat the words of a song twice and she would sing forth enthusiastically. Anton, however, determined her musical aptitude must be both nurtured and widened. He encouraged her singing, but began to think of other ways to develop her musicality.

By 1965 Anton realized his official research was not taking too much of his time, and he could devote more to his violin studies. He visited the local market and bought an old violin for the ridiculously low price of one RMB (about $3). He sacrificed this scratchy-toned old instrument, took it apart carefully, and studied its components, making copious notes. He compared his findings with Edward Heron-Allen's book on violin making.

He also began teaching violin to a friend called Wang ZhengJun, who later became a violin teacher in Changchun. In good China-fashion Anton did not charge for tuition, but Wang made Anton an offer: he would obtain quality wood for him. This was a valuable proposition. Wood used for making violins (called tonewood) must fulfil exacting specifications. It should be grown on the sunny side of a mountain, and only the sunny side of the tree used. The wood needs to dry naturally for at least six years, when the process of crystallization of the wood fibers begins, and the fibers become lighter and harder. The process is complete after 40 years, when it becomes best for violin making.

Anton had access to physics department X-Ray machines to check the quality of wood his student brought. X-ray was used experimentally to determine the age and quality of wood. A good violin maker does not need to use X-Ray to assess wood, but for experimental purposes Anton decided it was important to document the process. As his knowledge of violin making and supply of quality wood increased Anton conceived an ambitious scheme. He would make a quarter size instrument for NeeZi, and teach her to play.

From his reading he had discovered that wood used to construct violins was tuned, although he was not sure how. He learned tapping wood gave a distinctive tone frequency, unique to each piece of wood, called the *Eigen* frequency. While still a student he read that the old Cremona violins had this tap tone, but different researchers stated different values and methods for determining how it should be used. Initially he compared wood tones with a xylophone borrowed from the university orchestra. Then he made his own set of wood plates, and tried to identify their *Eigen* tap tone.

Anton did this early in the morning, before his official university duties. His neighbors called him "monk", because Buddhist monks tap their bowls as they beg. Some were less complimentary and called him "*guai jie*", meaning geek, but they were still friendly. For seven long years Anton persisted with his efforts to identify the *Eigen* frequency of various woods, until finally his ears were sufficiently tuned that he could accurately identify the tone of any piece of wood. Many years later he heard that an Italian luthier had developed the Luccimeter, named after him, that detected wood frequencies. But the cost of this device is high, and only large companies can afford to use it.

Eventually Anton concluded:

1. The best model for a violin is the Stradivarius flat model.
2. The top and back of the violin should have at least one matching *Eigen* frequency.
3. The best result occurs if there are two or more matching *Eigen* frequencies.
4. Wood with *Eigen* frequencies lying between tones D-E on the base of the violin results in a violin that is easy to play with a nice mellow tone. An E-F base tone violin sounds brighter and more powerful, with a strong penetrating tone, but it requires strong bowing to produce good playing results. An F-sharp base has the most powerful and brilliant tone, but it is very hard to play well.

Many years later, this method, especially points two and three, was recognized as Anton Sie Bitri Harmonic tone matching, critical factors required to make a good violin. Anton discovered, from experiments others made with old Italian violins, especially those attributed to Stradivarius, that the best woods for making a violin are believed to be Bosnian maple matched with Norwegian spruce. However, Anton successfully made excellent violins using Chinese woods.

He decided all this knowledge should be tested by making NeeZi a violin.

The quarter size violin he made for her was both a labor of love and an exposition of his scientific research into the secrets of making the best violins. The chisels made with Li the ironworker were vital, but at that time Anton lacked many other tools. He used broken glass as a scraper to painstakingly shave the wood and achieve the tone he wanted for the two plates of the tiny violin.

When he brought the violin home and gave it to NeeZi he felt well rewarded for his efforts. She danced around ecstatically, singing the latest modern Beijing opera arias at the top of her voice while pretending to play her violin. Each day after work Anton gently showed her where to place her fingers, how to pull the bow, and how to read music. As she progressed he got out his guitar and accompanied her simple tunes to make memorable music.

Unfortunately, NeeZi's excitement and enthusiasm for the violin did not last long. Anton decided that in the interests of music and science, as well as his daughter's long-term development, he must break his golden rule not to force someone to do something. He believed he was disciplining his daughter, not forcing her, and gently (and sometimes not so gently) urged her to persist with her violin practice. His method was simple. When he arrived home from work, he would ask if she had practiced. When he looked at her with his clear, penetrating gaze, she never dared tell a lie. If she had failed to practice, she was asked to do it after the family evening meal. NeeZi finally realized how important her violin practice was to her father, and settled down to regular effort. She noted he never asked her whether she had done her homework, but only about her violin practice. Practicing after the evening meal posed a problem for the family: with just one room for all activities, it was difficult for anyone to do anything else while she was playing. In summer YuXiang might go for a walk, but in winter this was impossible. NeeZi finally practiced faithfully without encouragement when she first got home from school.

When she was about six-years-old Anton felt her playing had progressed sufficiently to present her for examination by the Central Conservatory of Music. His little daughter performed well, and the examiners were very pleased. But what surprised them was this small girl played an instrument that sounded very good, not at all like the scratchy violins used by other children.

"Where did you get your violin from?" they asked when she had finished her exams.

"Oh, my Daddy made it, especially for me!" she declared proudly, and held it out for them to see.

"Did your daddy bring you here?" the examiners asked.

"Oh yes! He's waiting outside for me."

Anton was called in, and, with increasing pride and surprise, watched as the four examiners carefully scrutinized his daughter's instrument.

"This is a very good violin," they finally declared. "Did you really make it yourself?"

"Yes, I work at the physics research unit of Jilin University. One of my projects has been studying the acoustical qualities of good violins."

"Really? That is most interesting," the examiners declared solemnly, handing back the small violin.

Anton and NeeZi took the bus home and joyfully reported the day's events to YuXiang.

"Seems as though your ideas for making violins are good," YuXiang said encouragingly, although not without misgivings. Anton's violin making was a major problem for her. Wood shavings and dust invaded every part of their home, despite her rigorous efforts to keep things clean.

A few months later Professor Sun ChiaChung came to Anton's workstation.

"Well, well, well! Look at this!" he grinned. "Don't know how they heard about you, but Dr. Wang Xiang of the Central Musical Instrument Research Centre in Beijing wants you to go there and work with him." He drew out another red-chopped letter and spread it out before Anton.

Anton stood there, astounded. "But . . . but . . . my work here," he stammered.

"Nah, nah, don't worry! There's no way we would agree to release you, not unless the Chairman himself demands it!" Professor Sun laughed. "But I think I can work out a deal with Dr. Wang. It would be good for us to have a connection with his department."

It was finally agreed that Anton should go to Beijing for one week every three months. There he studied the physical properties of musical instruments. The first project he worked on was calculating the frequencies of *Bianzhong*, ancient Chinese "flat bells." They worked on the thousand-year-old set of bells found in the tomb of Ma Huang Thom. With Anton's unique aural ability to test musical tone, the study produced excellent results.

*****

Anton's musical interest was not confined to the study of the violin. He investigated his first instrument, the guitar, and concluded that not only was the guitar a powerful folk instrument, but it could be used effectively to play classic Chinese music. This led him to examine the Chinese *pipa*. He was able to show that the *pipa* had similar properties to a guitar. He amazed his friends by playing the *pipa*. Even though he had no formal training to play this instrument, he simply tuned it to match the guitar, and used the same fingering. The correspondence between guitar and *pipa* became recognized in the music world. A few years after Anton did his work, the author attended in Hong Kong a beautiful concert where *pipa* and guitar performers alternately played together and exchanged instruments.

Anton's interest in fine art was a major factor in his relationship with YuXiang. He developed his self-taught skills to a high degree, and others noticed this. During his years as Researcher in the physics department he was commissioned to paint at least ten large portraits of Mao Zedong. These portraits were in high demand, not only because of Anton's obvious skill, but also because they were, at the time, virtually the only decoration used in public squares or large government buildings.

Anton had an excellent relationship with the professor of his department. Professor Sun ChiaChung taught him many things that helped his research, especially the use of various mechanical and electronic calculators, the early forms of computer. He especially appreciated Professor Sun as a person. Professor Sun was modest, unassuming, and extremely honest. He often said the most important characteristic of a good scientist was honesty, and the willingness to study problems step by step without hurrying to get results. He advised that no one should fight for best and first place, because science was built on the discoveries of others who had gone before. Science is a relay race, where the baton is passed from one to another participant.

Professor Sun lived a simple life, and was willing to help his staff. He kept his bank account book on his desk, and offered that anyone in financial distress was free to draw on his account to deal with their problems. The professor was not only willing, but encouraged staff and pupils to succeed and overtake him in their achievements. For Anton, he became a powerful role model. Another member of the university staff who gave considerable financial support to others was Professor Xu ShaoHong.

When the Cultural Revolution gripped China and disrupted the academic life of students, it actually had a positive effect on Anton's research. The university was closed to students, but all staff continued their research. In fact, there was a strong central government directive that university science departments should continue without interruption. Since there were no students to take the time of staff, they proved to be more productive in their research. What was seen as a disaster for many people was Anton's chance to study the acoustic secrets of the violin makers of Cremona. He checked and rechecked the results of his meticulous experiments, until he was confident he understood how to make top quality violins. Then utilizing his knowledge, he made two violins that served as his personal teaching instruments for many years to come.

*****

But while Anton was content and achieving success in various aspects of art, science, and musical research, YuXiang was struggling to cope. Daily she became more haggard and discouraged.

In 1969 their second daughter YanYan was born. It caused a major disruption of their single room.

"I have no money for a bed for our new child," announced YuXiang dejectedly.

"Don't worry," Anton assured her. "I'll make something."

He dragged his precious aluminum cases from under the marital bed, found a sheet of wallboard, and laid it across them.

"We just need a mattress," he said, standing back to admire his efforts.

"Mother and I have made plenty of quilts," said YuXiang, brightening. She laid a pile of thickly padded quilts over the board, and three more lightly padded ones became coverlets.

Initially, while YuXiang was breast feeding YanYan, Anton slept on this bed-on-boxes, and YuXiang and the girls had the double bed. But soon six-year-old NeeZi was moved to the box bed. She became very attached to the aluminum boxes that not only provided her with a bed, but also housed her father's violin-making equipment.

Unfortunately, the joyous arrival of YanYan complicated YuXiang's deteriorating health. Adding to her problems was the fact that her parents, who had lived nearby and previously given her considerable support, moved into the country. The reason for their move was zealous patriotism. They were originally country people, and had lived in Changchun for only a few years. YuXiang herself was born in the village of Yong Ji near Changchun. When the hype and excitement of the Cultural Revolution began there was official government encouragement for people to move into the countryside to build up China. Although it is well known that students took up this challenge with enthusiasm, they were not the only ones. Mother and Father Wang thought that to help peasant farmers for a while was a noble idea. No one coerced them to move. Unfortunately, however, once out of the city they found themselves unable to return, and trapped by the restrictions of Chinese country living. Her parents' plight initiated in YuXiang a slow loss of faith in the country's system of government.

The Wangs' home in Changchun had been about an hour's walk from Anton's and YuXiang's apartment. But their new home in Shuang Yang county was thirty-five kilometers away, a distance too great for YuXiang or her children to walk. Their new home was similar to their Changchun house, traditional bungalow style with the *kang* forming the center of all activities, but there were significant differences: the floor of their Shuang Yang home was merely packed earth, there was no electricity, all water had to be carried from a well some distance away, and there was no flush toilet. Schooling and employment opportunities for YuXiang's two younger sisters became very limited. YuXiang was refined and gentle in her manner, but her two younger sisters, both good students, felt they had been duped, and became embittered. They frequently complained about their lot, and blamed their mother for dragging them from city comforts to suffer deprived country living, coerced against their wills to live where they least wanted to be. YuXiang's mother agreed the decision to move was not good, and deeply regretted her choice, but government regulations meant there was nothing she could do about it. The sisters' anger against YuXiang's apparently better situation compared to their predicament caused family stress. Fortunately, several years later both sisters had opportunity to further their education beyond the level YuXiang had achieved, and became successful professionals.

The thirty-five kilometer trip to her grandparents' home (almost always with her father) became a long ordeal for NeeZi every school holiday. Because YuXiang worked fulltime, she needed her parents to care for NeeZi during holidays. NeeZi's grandfather never took the bus to Changchun, and always walked to visit his daughter (taking a whole day each way), but NeeZi was too small to do that. The bus was old, dirty, and battered, its windows rattling and unglazed. It was hard even for adults to cope with travel in the severe winters. The roads were unsealed, and the bus stopped frequently, so the journey took at least three hours, often four. Even when they arrived at the Shuang Yang bus stop there were still two kilometers to walk down a muddy track to the grandparents' home, a walk NeeZi found a trial. If they were lucky there might be a donkey cart or horse dray to ride, but these were rare and drivers often refused to take passengers. When YanYan left the *danwei* crèche and went to school both NeeZi and YanYan went to their grandparents for school holidays.

But once at the Wang home the girls reveled in good food, such as they never ate in the city. One reason YuXiang was keen her daughters go to her parents' home was NeeZi was often sick. YuXiang believed she was malnourished, and her mother's food would help. Not only was grandmother's cornbread tasty, but there were plenty of eggs and pork from the chickens and pigs grandmother kept, and heaps of vegetables, fresh and preserved.

NeeZi loved her grandmother dearly, and grandfather was always kind and attentive. She knew she was special because when there on her own she was allowed to sleep beside her grandmother on the *kang*, even before her aunts who should have had priority. Normally older family members slept closest to the kitchen wall of the *kang*, where it was warmest, and brothers and sisters on the cooler outer edge. Guests came second in priority, and slept on the outermost edge, after the family.

Beloved Grandmother Wang, however, was more than merely kind. YanYan also loved staying with her grandparents. But whereas NeeZi played daintily with girls, YanYan played mud-pies with boys, and hunted for wild garlic and dandelions. It was especially good to find frogs, and take them home for grandma to cook. One day YanYan found a hen's nest containing three eggs on the wall of a neighbor's pigsty. She knew she should let the neighbors to find the nest, but convinced herself that they would not, and anyway, eggs were so special, such a rare treat, she could not resist them. Country people had plenty of eggs, she reasoned. She went away, but could not stop thinking about the eggs on the pigsty wall. Finally, she gave in, and collected them. Then in great excitement she showed her treasure to grandmother. But Grandmother Wang knew she had collected all her own eggs for the day.

"Where did you find these eggs?" she asked sternly.

"Oh, out there," responded YanYan, waving her hand vaguely.

"But where out there?"

"Oh, where I was playing." More twirling hand movements.

"Did you find these eggs in my house or the neighbor's?"

"On your wall." YanYan began skipping towards the door.

"Stop! You know very well we don't have a wall! The only wall here belongs to the Chens."

"Does it? Really?"

"It does, and you know it does!"

YanYan hung her head. She was beaten, and she knew it.

"Now take those eggs back to the Chens and tell them you are sorry!"

"Can't you do that, grandmother?"

"Did I take the eggs?"

YanYan took the eggs back, endured the severe scolding she received, and never forgot the lesson in honesty.

The village children loved playing with the two city girls, which made NeeZi and YanYan feel special. But when they returned to Changchun the sisters were always sure of one thing: a very long, very deep, and very thorough scrub in the bathtub.

"I've no idea how you get so dirty out there!" YuXiang grumbled as she plied a stiff and tickling brush to their arms and legs. The joy of the bath however, did not make either girl become too worried about keeping clean.

But while her daughters enjoyed improved health from grandmother's cooking, YuXiang struggled. Life for her was an unrelenting round of hard, demanding work. She was first up in the morning, and last to bed at night, but her work was never finished.

The family was provided with very limited food coupons. It was YuXiang's task to first obtain this limited food supply, and then make sure it would stretch from one issue till the next. In general, red sorghum, couscous and cornmeal were sufficiently available. Precious wheat flour was reserved for special treats such as dumplings for New Year celebrations, and rice was so precious it was generally used only when someone was sick. Meat coupons allowed for 500 grams per person per month of either meat or eggs, but not both. Soy bean oil, which supplied flavor as well as energy, was issued at 250 milliliters (about eight ounces) per month per person. It took significant ingenuity to transform these small quantities into tasty and nutritious meals.

Buying vegetables, a spring and summer activity, was a major challenge. YuXiang asked Anton to help with this chore as he worked close to the shops, but he was hopelessly unsuccessful. Anton was always polite, always willing to take his turn queuing for food, but he never joined the stampede, and never succeeded in buying a reliable supply of fruit or vegetables.

"But how can you say, 'I just can't push and shove'?" YuXiang would scold. "You can eat, can't you? You know I can't get to the shop! I work too far away. You've got to help!"

Anton would sigh, frown, and agree that he would try again next time, but he never did any better.

Very soon YuXiang had to admit he was absolutely no use buying food, or anything for that matter. She tried cajoling, she tried nagging, she tried demanding. But it was no use. Her talking made Anton angry, but did not change him. He remained a hopeless, helpless spectator in the battle for food. At first Grandmother Wang helped, but when she moved to the country NeeZi, as elder daughter and now about seven, like most other older Chinese children, was pressed into service.

The precious vegetables arrived, irregularly, in the afternoons when NeeZi was home from school. Someone in the apartment block would spy the horse-driven vegetable cart coming, and start the shout, "Vegetables are coming! Vegetables are coming!"

NeeZi instantly stopped whatever she was doing, raced home to get coupons and money, and sprinted to the shop. Fortunately, because the Sie family lived close to the vegetable shop, she often got there while the cart was still being unloaded and before other people arrived. She might even be lucky enough to buy the vegetables and get out of the shop before the crowd got there. This made her very happy because there was always shouting, arguments, and sometimes even fights as customers grabbed the precious vegetables. Within half an hour the shop would be completely empty, and there would be nothing more for several days.

In spring, the first vegetables available were spinach, leek, scallion, and cucumber. In summer, beans, *pakchoi*, tomato, eggplant, celery, and green peppers were added to the available list. From the erratic supply NeeZi obtained YuXiang tried to devise healthy and balanced meals. It was no easy chore. The supply of summer fruit was just as erratic and limited. Although there was a good variety that included watermelon, pears, apples, apricots, plums, persimmon, and hawthorn berries (YanYan's favorite), available quantities were seriously inadequate.

As well as the challenge of feeding her family, YuXiang had to clothe them. For this she was also issued coupons, but the supply, as for food, was extremely limited. For summer her children had only two or three tops, one skirt and a couple of pairs of trousers and these had to be carefully recycled. The local saying was: "Three years new, three years old, mend it, patch it, three years more." Everything was mended and recycled. As the weather became cooler in late summer a knitted sweater and knitted trousers (of course made by YuXiang) were added to the wardrobe. Then in winter thick three-layered cotton-padded jackets and trousers were added, with finally, for outdoors, the hooded jacket called a "monkey" topped it all. In Changchun those winter clothes had to last six months. Washing clothes in the extreme cold was virtually impossible.

But every year YuXiang, despite her fulltime work, unpicked outgrown woolen garments and re-knitted new sweaters to fit her daughters, tediously matching and joining worn pieces of wool to make whole "new" garments. NeeZi and YanYan, like all Chinese children, especially girls, learned to knit their own scarves, hats, socks, and gloves at a young age. Occasionally there were enough coupons and money to buy new yarn to make a sweater, or new fabric for clothes. What excitement!

Every year, at winter's end, the padded clothes had to be unpicked by YuXiang, opened up, and all parts washed, including the thick central layer of cotton wadding. Then, entirely by hand because they owned no sewing machine, YuXiang would remake these essential three-layered winter garments. If any of the fabric was too worn to be reused for clothing it was recycled as shoes.

Grandmother Wang took responsibility for making shoes for the family. Both uppers and soles were made from worn out clothes, layers and layers glued and stitched together with strong hemp thread. The soles were so strong that when newly made NeeZi could stand on her toes, and swirl around pretending to be a ballet dancer. Grandmother made thick winter shoes and light summer shoes, but unfortunately neither was waterproof. Walking along the snow-covered roads in winter was a painful hazard. There were no road sweepers clearing the snow from road or footpath, and snow could be a meter high.

Most people could afford only one set of padded clothing. Adults took great care of these greatly treasured garments. They knew their survival depended on them. But children did not understand. One day NeeZi was too busy having fun playing with her friends after school to respond to signals from her bladder. Her one and only pair of padded trousers was soaked with urine. But in the icy cold weather there was no way YuXiang could get them washed and dried. The pants were spread out on the heater that was supplied with hot water a mere two hours each morning and evening. NeeZi had to use those smelly pants till next spring. She never made that mistake again.

Water was also in short supply. The apartment had the luxury of a water pipe in the kitchen, but water flowed from it only one or two hours every morning and evening. Although all the family was supposed to help, it was YuXiang's responsibility to make sure the large tank in the kitchen was filled with water. Once stored the precious water had to be rationed carefully. Baths, even in a small tub, were a rarity, and, as noted, washing winter clothes a yearly luxury. Initially there was a *danwei* public bathhouse for the family to use, but as supplies of water, and coal for heating became scarce, this benefit ceased.

Another family task, one the girls enjoyed although their parents did not, became YuXiang's responsibility: to ensure windows were properly air-proofed for winter. The double-glazed windows had strips of paper glued around all edges inside and out. YuXiang made five-centimeter strips from old newspapers, and glue from precious wheat flour. NeeZi's job was to paste glue on the strips. It was Anton's job to apply the sticky strips to the windows, the hardest part reaching the outside of the small "ventilation" pane at the top center of the window. Anton did not like doing this, partly because his mind was focused on his research, but also because it meant leaning far out from the other second floor windows to reach the ventilation pane. Winter brought one good thing: the cold made beautiful ice patterns on the windowpanes, and these patterns were different every day. People whose apartments faced north, however, did not have the pleasure of changing ice patterns; theirs remained the same all winter because there was no sun to melt them.

It was no surprise that work, mothering, and coping with China's many shortages, without any labor saving devices or the nearby support of her family, took its toll on YuXiang. About 1970, while YanYan was still a baby, YuXiang's chronic cough began producing blood-stained mucus. She was diagnosed with tuberculosis. For months she underwent treatment, with daily medication that made her nauseated, and long weekly trips to the hospital for injections. But the blood stained mucus continued. She lost weight and struggled to cope. There seemed to be no light at the end of her bleak tunnel of discouragement.

Anton, very worried, took her to the Xiehe hospital in Beijing on one of his trips to the Beijing Central Musical Instrument Research Centre. Extensive testing did not reveal the presence of any current tuberculosis bacilli, but the bleeding continued. Doctors at this famous hospital declared the problem was caused by bleeding capillaries in YuXiang's throat, but they believed her general health was not good enough to contemplate surgery to correct the problem. Anton was frustrated by these verdicts, and after the long trip YuXiang returned to Changchun utterly exhausted and even more discouraged.

Like many people in China, they turned to traditional Chinese medicine. Anton practiced acupuncture on his wife, prepared many herbal concoctions, and did his best to treat her illness, but all to no avail. She remained sick and disheartened.

*****

Although he was unable to help his wife, Anton's excursion into barefoot doctor role brought spectacular relief to at least three people. Huang You-Bao, dean of the Science faculty, had a stroke, and was unable to walk. Dean Huang was greatly loved by his staff. Anton studied his acupuncture books carefully, and began to puncture his Dean. To everyone's delight after one month Dean Huang was able to walk again. This success meant other faculty brought their sick family members to Anton for his help.

On two occasions Anton helped faculty whose wives suffered from mania. Both women were brought to him accompanied by four men and roped up like pigs so they could not escape. Anton used twelve puncture points, as recommended, to help the first woman, wife of Liu GuoLu, and was astonished to find that when he pierced the second finger nail as instructed she suddenly calmed down and began to speak normally. He was able to visit the grateful Mrs. Liu about twenty years later, and she had remained well during the intervening years. The other woman was the wife of Sun YuHang, and her story and response were similar.

Understandably, after these three successes, Anton became known for his acupuncture skills, and was in demand by staff members and their families, even though he could not help YuXiang.

*****

Despite difficulties, there were many small enjoyments in life. No one had vacations, and no one expected them. But there were happy family activities and outings.

The shortage of storage space goaded families in apartment block 2200 to make cellars in the courtyard of their L-shaped building. These cellars enabled families to store enough vegetables to last through winter (without blocking the stairs), and their construction provided endless enjoyment for the children. They loved helping their fathers, and what did it matter if they got covered in mud? It was a wonderful excuse for a precious bath! Stocking the cellar also provided huge fun. The children climbed down into the deep muddy hole squealing, sliding and laughing, while parents handed down vegetables. Cabbages had to be stacked carefully so as not to fall. The dry leaves of radishes and peppers were kicked around with great glee. The soil removed to make the cellars became a small hill in the middle of the courtyard. When snow-covered this was a wonderful tobogganing area. But in spring and early summer, until covered with grass and weeds, it was a horrible mud heap, and a nightmare for mothers with rationed water.

At Chinese New Year there were public fireworks, exciting communal celebrations. Personal celebrations such as birthdays never occurred, but National Day (October 1) and New Year provided pleasure for everyone. Despite the hazards of getting lost in the crowds, or worse, getting trampled on as winter roads were icy and it was easy to fall, everyone loved these events. The display was always staged from Peoples' Square in the center of Stalin Avenue (which became Peoples' Avenue after the break with Russia). But the Sie family had an accident proof venue: their own balcony provided the perfect place for viewing the fireworks.

The small food shop close to their apartment sold a variety of goods, if you had coupons or money to buy. Candles, biscuits, sugar, soya sauce, vinegar and so on were available. NeeZi ardently wished whenever she went to the shop she could persuade someone to buy green bean cakes, her greatest enjoyment. Years later she was delighted to return to Changchun and find her beloved green bean cakes still for sale in this same little shop.

Seated outside the shop, in both bright summer sun and icy winter cold, was an old woman with a wooden trolley loaded with a large blue wooden box. She sold small "ice creams" on sticks. These were just sugar and water, and certainly no cream! As a rare treat someone would buy the Sie girls an ice. No coupons were needed for these, and if you had enough money you could even buy a red or green bean ice. Although rather starchy they did not melt so quickly.

Walks in nearby parks during summer were fun. No one had pets, but people improvised: take the family hen for a walk. Once YuXiang and

YanYan took one of their hens to the park. This hen did not lay well, but the family was fond of her, still fed her, and enjoyed letting her run on the grass. Suddenly YuXiang noticed the hen's behavior was odd, and feeling the bird's abdomen realized it was about to lay one of its rare eggs. What wonderful news!

"Hurry, YanYan!" she admonished, carefully picking up the hen and cradling it in her arms. "Our hen is going to give us an egg! We must get home quick!"

Faster and faster they hurried. Unexpectedly YuXiang tripped on a rock and fell to the path. Ignoring her own scratches, she got up clutching the precious hen more tightly. Suddenly mother and daughter heard a most disheartening "Cruckkk!"

Looking down, they saw to their utter dismay that the stressed (and pressed) hen had indeed laid her egg. There spread over the dirty ground was a sticky yellow mess.

"Oh no!" wailed YanYan. "Now we can't have egg custard!"

Savory egg custards were one of her favorite foods.

"Nor will there be any sweet egg roll," thought YuXiang, her favorite treat, and something Anton loved.

About 1974 the Physics Institute was allocated new land next to beautiful South Lake. New laboratories were built, new dormitories and a new canteen. The Sie family was directed to move to this highly desirable location, and were very happy. Their apartment on the sixth floor now had wonderful views, their own kitchen, their own toilet, two bedrooms, and two living rooms for them to spread around. What luxury! Food also began to be more plentiful at this time.

A particularly good aspect of this new apartment was nearby South Lake Park. It was a beautiful place for walking, especially so in autumn when it had an abundance of wild fruits and mushrooms free for anyone to gather. YanYan was always keen to gather hawthorn berries, but because they grew high on trees she needed help to collect them. What delight when she could take her treasure trove home!

Mushrooms were another great treat. One sunny autumn day five-year-old YanYan and YuXiang were in South Lake Park when YanYan found an unusually large white mushroom. Mushrooms were never easy to find, but one this big was rare indeed. YuXiang helped her daughter carefully place the precious mushroom in her bag, and they took it home to cook as a treat for their evening meal. Inexplicably, YanYan decided to place the treasured mushroom on the windowsill where she believed it would keep in perfect condition ready for cooking.

Later that afternoon YanYan noticed the paper under the mushroom had moved. Climbing on the bench to investigate, she found to her utter dismay the mushroom had gone.

"Mother! Mother! Mother! My mushroom!" she began wailing. "It's gone!"

YuXiang came to investigate, and realized the wind had blown the mushroom out the window. They climbed down six floors of stairs and looked hard, but only a few broken pieces of mushroom could be found. YanYan was heartbroken! It took a long time for YuXiang to comfort her small daughter. But, although they never enjoyed this mushroom treat, it remained a precious memory of what South Lake Park could provide.

One day Anton arrived home with a small package containing a strange-smelling yellow substance he said he found in the small local food store. The family gathered around, mystified.

"What is it, Father?" asked YanYan.

"It smells very odd," observed NeeZi, wrinkling her nose.

"It's butter. No one else wanted to buy it because they didn't know what it was. But I remember it from when I lived in Semarang. Dutch people love butter."

"Butter? But it smells very strange. Is it really good to eat?" responded NeeZi doubtfully.

Anton opened a larger package bought from the shop.

"Oooh! *Mantou!*" squealed YanYan.

Anton smiled as he placed the Chinese bread rolls on a platter. He reached for the small jar of sugar on the top shelf behind him, spread some butter on the *mantou* and topped it with a sprinkle of sugar. He cut one *mantou* into four parts, and handed a small piece to his wife and daughters.

"This is heaven!" declared YanYan, her mouth full of the delicious new food. The family laughed, agreed, and reached for another portion of this rare treat.

"Father," asked NeeZi, thoughtfully chewing on her delicious buttered and sugared *mantou*. "What other things did you have to eat in Semarang?"

"Oh, we had lots of fruit. Bananas and . . ."

"Bananas! Did you *really* eat bananas?"

"Yes, we had a banana tree in our garden. I had banana for breakfast every day."

"Every day?"

"Yes."

"Then you really did live in heaven!" declared YanYan solemnly.

Anton roared with laughter.

YuXiang looked very thoughtful.

*****

One icy winter's day late in 1975 Anton arrived home earlier than usual, waving a letter from overseas. Letters were rare; during the years 1966 to 1970 the family received no letters at all.

"Look at this," he said to YuXiang.

"It's from your family," she responded, looking at the stamp on the envelope. "What does it say?" YuXiang could not read Indonesian.

"My father says my mother is sick. He wants me to see her, although he doesn't quite say that."

"But how? You've told me so often you can't go back to Indonesia!"

"Maybe there is a way. Perhaps I had better discuss this with my *danwei*."

YuXiang shrugged. Couldn't Anton accept the fact that his education choices meant he was trapped in China, just as her mother was stuck in Shuang Yang? Perhaps he knew something he had not told her. Perhaps, even if he could not go to Indonesia, he could leave China. Perhaps, just perhaps, she too could begin a new life.

Anton took the letter to Dean Huang who agreed that because Anton was an overseas Chinese he was free to leave China at any time. Like all Chinese Dean Huang recognized the importance of family ties, and family duty. If Anton's mother was sick, then he should see her. The complication for Anton was therefore not the Chinese government, but the Indonesian, which still had a negative attitude to China. Anton's Chinese education barred him from returning to his homeland. It was not until 1980 that the Indonesian government relaxed its regulations and allowed Indonesian-born Chinese who had entered the People's Republic of China to visit their homeland, although they were still denied the right to hold Indonesian citizenship.

"Why don't you go to Hong Kong?" Dean Huang suggested. "That could be arranged quite easily. Get your parents to go there, and you can see them."

When Anton shared this idea with YuXiang she opposed it, and adamantly declared, "We have lived together for more than twelve years. We should die together."

"I'm not talking about dying!" he said roughly.

"But if you leave here, even for a short time, they might make our children and me leave this house! They might make us go back to my parents' village! I struggle to live with my wage and yours. How could I ever support myself and our daughters with just mine!"

Anton knew YuXiang was far from well. If he left her to go to Hong Kong, even for a visit, her future could indeed be very insecure. Their apartment was allocated from his *danwei*, not hers. She was right, it was possible she could be forced to move out of her home. He admitted he had no idea how long he would be in Hong Kong, and indeed, if once he left China whether he could return and resume his present job.

Anton went back to Dean Huang.

"You could try going to the Philippines and get a Philippine visa, and then use that to visit your parents in Indonesia," he suggested when Anton told him YuXiang was not happy about the Hong Kong plan.

But this idea was even less acceptable to YuXiang. "That's crazy! How long do you think it would take to get a Filipino passport?" she demanded.

Anton had no idea. Preoccupied with his work, for several months he pondered his options.

"Well, it could be possible for all your family to go to Hong Kong," said the Dean, doubtfully, when Anton finally returned to him. "Hong Kong is very tough on illegal immigrants these days, and they don't seem too friendly towards immigrants from the mainland, as they call people from China. But it might be possible. I mean, the government would give your Indonesian passport back."

When Anton shared this with YuXiang to his surprise she was very enthusiastic. While he had been absorbed in his research, YuXiang had been dreaming of distant countries, of the possibility that somehow her burdens could be eased, and her health improved.

"But I would never be able to get work as a scientist in Hong Kong," said Anton gloomily. "I enjoy my work here, and life here is definitely improving. We have a good apartment now, and food is much more freely available. Why don't I just visit Hong Kong for a few days and get my parents to visit there at the same time."

"No!" repeated YuXiang vehemently. "I don't want to be left here alone with two small children! I don't know what would happen to us when you are gone! As a family, we should stay together. And anyway, look at what else you can do! I'm sure you could get work!" she countered. "There's your work making violins. You could teach violin, or do your painting. Then you do very well with your acupuncture. Perhaps you could be a traditional Chinese doctor in Hong Kong. You are good at so many things. You could start by doing carpentering work; surely that would be easy to find. I've heard that carpenters get much more money in Hong Kong than scientists get here in China. I've heard there's a lot of building going on in Hong Kong. And think of our girls, of their future and opportunities."

"Perhaps," said Anton, and turned away, adding bitterly to himself, "but what would she know about the rest of the world?" Suddenly he realized he did not want to leave China. He did not even want to think about it. How could YuXiang even consider such an idea? What had given her such crazy thoughts? How had she gained information about Hong Kong when he knew nothing about it?

YuXiang broke into his reverie. "It's alright for you, you have a good job. People respect you. They treat you well. But . . . for me life is tough! It couldn't possibly be worse somewhere else!"

"What are you talking about?" he demanded.

"Well, at least it would be warm in Hong Kong. I might get better."

Anton dared not say a word. She just might be right.

YuXiang saw her life stretching out as an unrelenting struggle, a round of endless work. This Hong Kong idea gave her a glimmer of hope, and she was very persistent. Eagerly she gleaned every piece of information she could about Hong Kong. Anton also gleaned information. He learned that Zhou EnLai had signed a special paper facilitating the exit of overseas Chinese families. YuXiang's wish to leave might be easier to fulfil than he thought.

While Anton dallied and discussed his problem with his friends, 1976 ushered in a very tough year for China. First, on the 8th of January, the country was devastated by the death of Zhou Enlai, the man who had oiled the complex machinery of government for so many years. Endlessly the public radios broadcast sad, solemn music, and a heavy gloom settled on the nation. Even children as young as YanYan recognized the deep sadness and solemnity of the situation.

Then in the early hours of July 27 a massive earthquake struck the city of Tangshan, and at least a quarter of a million people died. The quake caused damage in Beijing and Tianjin, and was felt in Changchun, 750 kilometers away. Severe natural disasters in China were always given significant political implications, and were commonly seen as heralding the end of a dynasty.

When, on the 9th of September, Mao Zedong died, Anton finally made up his mind.

"We'll go," he announced to YuXiang, who made no attempt to hide her relief and delight.

Once more Dean Huang was approached, and he gave Anton all the assistance needed. The science institute staff made the many essential applications, and gave Anton and his family excellent reports. It took only three months for all formalities to be completed. China gave formal permission for Sie Anton, scientist, and his wife Wang YuXiang, expert technical draughtsperson, and their two daughters, dependent school pupils, to leave the People's Republic of China. And the Royal Government of the British colony of Hong Kong gave formal permission for this same Sie Anton, scientist, with his family as described, to legally enter and take up residence in Hong Kong.

There was little for the family to pack. There were Anton's violin making tools, and the three violins he had made in China (counting the small one made for NeeZi and which now belonged to YanYan). The beautiful German-made violin Anton brought with him from Indonesia was sold to provide money when they reached Hong Kong. It was the only commercially valuable thing Anton owned, and he knew they would be in desperate need of financial resources when they got to Hong Kong. The grateful purchaser was the Changchun Film Factory Orchestra, with whom Anton had sometimes played. Everything the family owned was packed into the same two aluminum cases that NeeZi slept on, and Anton had brought with him to China twenty-two years before. He had no more material possessions now than when he came to China, but he came as one person and left as four.

In the aching cold and gloom of early January 1977 Anton closed the door on their apartment, and the family walked down the stairs to the truck (already loaded with the precious aluminum cases) the Science faculty had provided to take them to the railway station. Anton's heart was heavy, very heavy, and his shoulders drooped at the thought of the huge and unknown responsibilities that awaited him.

They were driven in semi darkness through the familiar snow-covered streets. NeeZi and YanYan's eyes were big with uncertainly. Only YuXiang seemed bright and cheerful, confident that things would work out. As they came in view of the station she was the first to see and comment on the new diesel engine and its carriages proudly standing at the platform.

When they arrived at the station Anton was surprised to see four large trucks parked outside. They had a festive air, waving banners that announced they were from Jilin University. Suddenly he realized the banners were talking about him, about bidding farewell to good Comrade Sie!

Each truck was packed with people, meaning there were no less than two hundred. Two hundred people, jostling to shake his hand, to hug him farewell, to sing "The East is Red" for him! Many were weeping. Many tried to say how much they appreciated him. Many just grabbed his hand and held it. Anton simply could not believe this outpouring of appreciation for him.

He tried to speak to them all, to thank them, to bid them farewell. He tried to say nice things, sensible things, memorable things about the great Chinese Fatherland. But he simply could not.

For the first time in his adult life, Sie Anton wept publically.

As the family settled into their comfortable seats on the train, and waved to the forest of arms fluttering outside the windows, Anton was oppressed with dark forebodings about his family's future. Now it all depended on him. In China he had his dream occupation, and a secure, albeit basic, livelihood. As the train gathered speed he had dreary visions of spending the rest of his life in Hong Kong as a construction worker. But filial duty was the core ideal he had grown up with, and he would not deny his parents their right to his support when they needed it.

Fortunately, as the train gathered speed and left the cold, but familiar sights of Changchun, he had no idea just how hard his future would to be.

Passengers aboard the *Tjiwangi*, en route from Jakarta to Hong Kong, June 1955.

Physics class, Changchun, China 1957. Anton is second from left in the back row, and Tang ZhiXiu is squatting in front of him, in the slightly darker jacket.

Anton accompanying violinist Sun Xin on his guitar, 1958, in the garden of Jilin (North East Peoples) University.

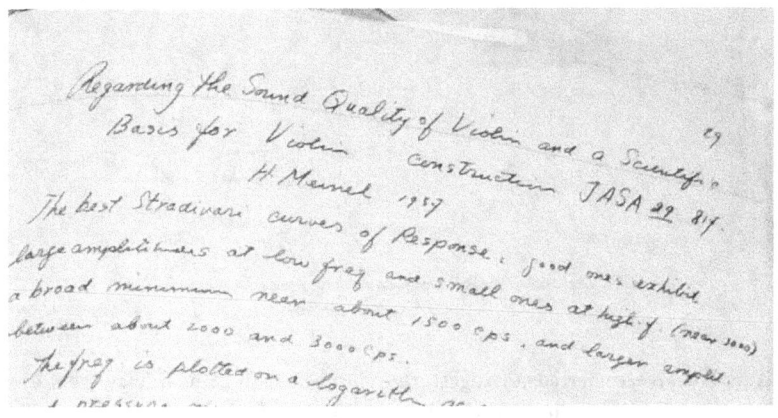

Detail of one of Anton's workbooks. Note, interestingly, they are written in English, indicating the quality of his training in Semarang, and perhaps his desire for privacy.

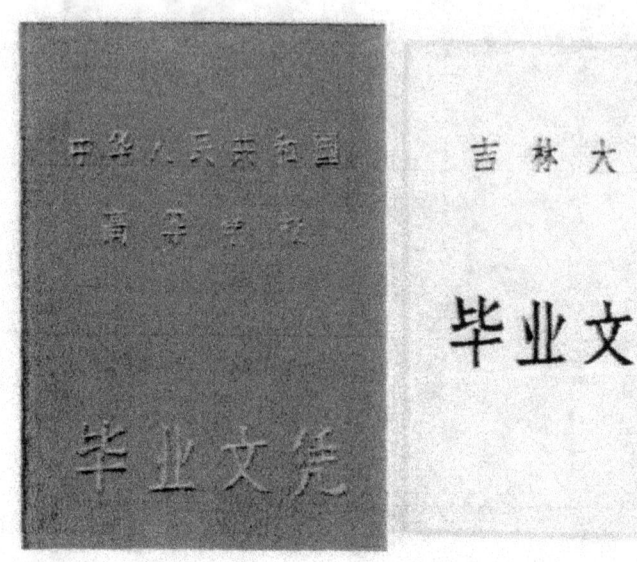

Kirin University Physical Department, Theoretical Physic Specialist Certificate of Graduation, 5-8-1961.

Anton's degree, conferred 5th August, 1961, just before his 26th birthday. It entitled him to be called "doctor" in Changchun.

Newly married Sie Anton and Wang YuXiang about 1963.

Indonesian Cultural Delegation, 1964. Anton has the beaming smile, front row second from R.

A 1967 tinted black and white studio photo of YuXiang, displaying Anton's skill at coloring.

Anton in 1969 with one of the ten large portraits of Mao Zedong he was commissioned to paint.

Sie Anton with his two daughters, NeeZi standing behind, and YanYan beside him, 1976.

Third Movement

Hong Kong Blues

Anton slipped quietly out of bed, carefully feeling his way in the dark. Gently he pushed open the door, its creaks shattering his wish not to disturb YuXiang in the lower wooden bunk and NeeZi and YanYan in the ancient iron bunk bed pushed hard against the opposite wall. The tiny apartment in Yuen Long had one bedroom and a living area, and was very small compared with the apartment they had left at South Lake, Changchun.

He stopped. The regularity of the family breathing did not change. Quietly he pulled the door behind him, and tiptoed into the living area. He felt his way through the gloom to the chair where hung clothes prudently prepared the night before. Mechanically he began to change.

Suddenly he felt a presence in the room, and froze. Hardly daring to breathe, he slowly turned to face the intruder. But it was only YuXiang, quietly lifting a kettle of water to their tiny two-ring gas stove. He sighed with relief.

"Here, I have prepared *mantou* for you," she whispered, pushing a small paper parcel towards him. "You will need energy."

Anton nodded gratefully, and sat down. YuXiang poured a cup of weak tea for her husband, and sat beside him.

"You should be asleep," he whispered. "It's not five o'clock yet."

"You have a long trip," YuXiang murmured. "Are you sure you know the way?"

"Yes, I have memorized the plan. First, number 68 bus to Tsuen Wan Ferry, then the ferry to the Island. Then I have the map Hung JiaDie made for me. See, in my pocket! I will follow it carefully."

Anton pulled a pack of cigarettes from his pocket, and lit one. YuXiang said nothing, but this was a sign he was worried. He was never a heavy smoker, but since arriving in Hong Kong he smoked very little, to conserve their extremely meagre supply of money.

"Better go," he said. "After all, this is why we came here."

"Yes, I hope the immigration officials will be understanding." YuXiang touched his arm gently.

Anton pulled his padded cotton jacket around him, and walked into the cold, dark night. He remembered Hong Kong as hot and steamy, like his Indonesian homeland, but in this mid-January pre-dawn it was very cold. An icy wind straight from the snow-covered steppes of Siberia clutched him when he opened the door. The bus terminus was mercifully close to their miniature apartment in Lian Sheng Mansion, on the main street, and he joined the queue of roughly clad men headed for construction work in the city. Suddenly he realized his worn cotton jacket was very different from the puffy, brightly-colored, waterproof jackets other men were wearing. He had made this journey three times over the past three days, but never so early. Most of the men in the queue, like him, were eating breakfast on the move. A soft drizzle began to fall. Men in the queue shivered, pulled up their collars and hoods, and covered their heads with the newspapers they had been reading. Anton had no newspaper or hood (he had given his monkey jacket to a friend in Beijing), but the other commuters ignored him.

Finally, the smartly brown-uniformed Kowloon Motor Bus Company driver arrived and unlocked the rear door of the bus. He took his time to arrange himself on his seat. Outside the queue shivered in the drizzle. At last the front door hissed open, and the passengers headed for seats and warmth. Anton, last on the bus, proffered the exact fare. He had learned painfully, on his first trip, that unlike the human conductors on Chinese buses, the mechanical slot machine of a Hong Kong bus never gave change. He chose a front seat beside a window, but as he lowered himself into it, the bus lurched forward. Anton fell heavily on the stocky young man beside him, who scowled.

At first it was impossible to see anything in the foggy gloom outside. With no sights to distract him, Anton mulled over the past few weeks.

He remembered little of the train trip from Changchun to Beijing. Perhaps, he reflected, it was just too familiar. He had, after all, made the journey many times to the Central Musical Instrument Research Centre. On those journeys he enjoyed watching the scenery; there was always something new. Perhaps, he thought wryly as the bus jostled him, his lack of memory was due to embarrassment at his loss of emotional control when he left Changchun. After all, it was not appropriate that a grown man should weep in public. But then he recollected many people had wept at the Changchun station: even Professor Sun and Dean Huang were weeping. Ah, it was so good to see Dean Huang walking. Anton smiled at the memory, and wondered how he had made the decision to say goodbye to his rewarding research career and all those wonderful comrades. Well, he did know:

filial duty to his parents, of course. Just what any elder son would do. The pain of leaving had eased when the Sie family was joined by another family also moving to Hong Kong. Chan WaiMan had worked for the Jilin Dancing Performance team, and Anton knew him slightly.

Both families spent a night in Beijing, in comfortable accommodation arranged by the Office of Overseas Chinese. Anton remembered with joy his work at the Beijing Central Music Conservatory, and was delighted when at this parting visit his colleague Chen ZhiMing presented him with two valuable books (of great worth in the ensuing years), *Violins and Violinists* by Franz Farga, and *Italiansche Geigenbauer (Italian Violin Makers)* by Carl Jalovec. Professor Chen was a world expert on folk musical instruments and folk music. At the Conservatory Anton met his friend Tseng XiangHo. Mr. Tseng, one of the examiners who had noticed NeeZi's superior violin, was responsible for arranging his involvement with acoustical studies in Beijing. The support of the Office of Overseas Chinese and meeting these old friends helped dispel some of his dread of what lay ahead.

The journey from Beijing to Guangzhou was enjoyable. The two families had comfortable cabins, and at regular intervals peddlers walked along the corridors with trolleys of food. To keep the children happy, Anton and Chan WaiMan allowed them whatever they wished to eat. Anton pointed out sights to his daughters, such as Zhengzhou where he had eaten the memorable meal of chicken, and of course Changsha, near where Chairman Mao was born. He wondered how he had managed to sleep through most of such an interesting journey! This time he was wide-awake during the crossing of the mighty Yangtze River. Proudly Anton told his daughters about Chairman Mao's first swim in the Yangtze in 1956, soon after he, Anton, had arrived in China. Then, warming to his topic, he told them that Chairman Mao repeated this impressive feat in 1966, when as a seventy-two year-old he had swum in the mighty river for more than one hour. Swimming was not a feature of cold Changchun life, and NeeZi and YanYan were suitably impressed. Both girls knew what everyone in China knew, that the Yangtze was a very dangerous and treacherous river, and only very brave or very competent swimmers ever swam in its waters.

The long nights on the train gave Anton ample time to think. He decided his dread of becoming a construction worker in Hong Kong was not an inevitability. He could choose what he made of his life, even in Hong Kong. After all, he had made the choice to go to China to become a scientist. Whilst his chances of continuing as a scientist, or a science teacher, were not great, he could develop his increasing interest in music. The more he thought about this the more he looked to the future with hope. Initially he would work at anything available, as long as it did not interfere with his chance of developing a musical career. He was confident he understood the construction of a good violin, and could use this knowledge to repair or make them. He was well aware that this was risky, but twenty-two years before he had taken a risk going to China, and he was willing to try once more to follow his dreams. The books his friends at Beijing Central Music Conservatory had given him confirmed this direction.

After three days and two nights on the train they arrived in Guangzhou. How good it felt to stretch their legs! And how good to meet his old friend Professor Chen MingJun and his wife Wu ChongXiu. They had so much in common. Both men were from Indonesia and had studied in Jilin University. Both wives were from northern China. Professor Chen, a mathematician, went to China a few years before Anton, and was now assigned to ZhongShan University near Guangzhou. Straight from the biting cold of northern China the Sie family found Guangzhou hot and humid, and the tiny apartment Dr. Chen procured for their visit not exactly palatial. But the girls were fascinated by the green grass and myriads of insects on the university campus. They spent many happy hours collecting specimens with Auntie Wu. YanYan was so interested in these brightly colored creatures that Auntie Wu was convinced she would become a biologist. The Sie family was sorry when the time came for them to take the train to Hong Kong.

When they returned to Guangzhou station they were delighted to find the Chan family waiting for them. They too had used the stopover in Guangzhou to meet old friends. When the train to Shenzhen arrived it was as old and battered as the one Anton remembered from many years before. Then he was an optimistic nineteen-year-old full of the spirit of adventure. Now he felt burdened far beyond his years. Soon they would be at the end of the line, and then what?

Three hours after they left Canton, with a surge of intense emotion, Anton led his family across the Sham Chun River Bridge that he had crossed twenty-two years earlier. They were in Hong Kong! But with no bike, and a wife and daughters to help carry the two aluminum boxes, it did not seem the ordeal it had been so many years before.

However, this time there was no self-important Chinese official dealing with the mysteries of immigration, and things did not go smoothly for the Sie family. Mr. Chan and his family quickly passed the English Immigration checkpoint, but unfortunately the Chinese British officers, clad in impressively smart navy uniforms, were very suspicious of Anton's box of violin-making tools. He was taken aside, first to one room for questioning, then another. How was it, they demanded, that a supposed physicist with recommendations from the Jilin University Science Academy needed to bring these heavy carpentering tools and lumps of old wood into Hong Kong?

Patiently for two hours Anton answered their questions, and signed numerous affidavits. He showed them the two unique violins he had made himself: one had a lion head, and the other a dragon head. But his story generated more suspicion than understanding. Although he was prepared for interrogation, and eventually satisfied the officials, it was a grueling experience. The frightened faces of his wife and daughters added to his apprehension. Finally, his interrogators ran out of questions, and the Sie family passports were stamped. Wearily they gathered their meagre possessions and walked into capitalistic freedom.

To his joy and utter amazement, as well as intense relief, the Chan family was patiently waiting for them beside the Lo Wu station ticket counter, together with a total stranger. He proved to be a Chan relative willing to guide them into Hong Kong. Anton had prepared a crisp red Hong Kong one-hundred-dollar bill to pay for their fares on the Kowloon-Canton Railway, but the Chan's kindly relative waved his money aside, generously paying for these total strangers. Over the next few years Anton was amazed at how much help "mainland" Chinese in Hong Kong gave each other.

At Hung Hom station, the cavernous Tsim Sha Tsui terminus of the Kowloon Canton Railway, Huang JiaDie, a man Anton had known slightly in Changchun, met the Sie family and took them by taxi to his home in Tsuen Wan. The eyes of NeeZi and YanYan were agog at the density of motor traffic in Hong Kong.

YuXiang suppressed a murmur of shock when she saw the tiny Huang home, but gratefully accepted a lower bunk with Anton while the girls squeezed into the upper bunk. She well knew the generous Huang family was similarly squashed in their own beds in the other bedroom of their home. Mr. Huang's wife had prepared a delicious meal for the travelers. NeeZi's and YanYan's eyes grew wide with amazement at the sight of bowls of precious fluffy white rice, a large platter of chicken breasts, and various dishes piled with meat and vegetables. But when their hostess brought out a bowl of yellow bananas, green pears and bright oranges they could not restrain their gasps of delight. Anton's anxiety melted under the benign influence of these bountiful provisions.

"I've brought a friend to meet you," Mr. Huang announced the following evening, as he ushered in a guest. "Liau RiSheng is a musician, so he might help your interest in music. He works in one of the New Towns. He can give you some advice. Oh, he comes from Indonesia, like you."

Anton beamed with joy. "I come from Bandung," said Mr. Liau, in Indonesian.

"And I from Kudus! We're both from Java!" Anton grinned, joyfully reciprocating in his native tongue.

After happily exchanging stories about their boyhood in Java, Mr. Liau advised Anton to move to the town of Yuen Long. "You will find living expenses very much cheaper there, and there's plenty of construction work in the New Territories. Do you speak any Cantonese? No? Hmm. That is a problem. That will make it hard for you."

Anton cringed at the suggestion of "plenty of construction work." He did not want to start down the road of fulfilling his worst nightmare!

"But he can do many things," observed Mr. Huang, knowing Anton's horror of "construction work."

"My husband can paint. Music. Carpenter. Teach. He can do lots of things," added YuXiang.

"Really? Good. Well, you should be able to find something," said Mr. Liau crisply, realizing Anton had some clear ideas about his future. "But first, would you like to come with me tomorrow? I can show you a flat you could rent." His eyes glanced around the tiny apartment, quickly appraising the family's urgent need of accommodation.

YuXiang nodded enthusiastically.

Next day YuXiang again suppressed surprise, even horror, when she saw the flat offered them in Lian Sheng Mansion on the main road of Yuen Long. Their modest Changchun apartment was palatial in comparison with this tiny, rundown place. What had happened to the golden stories she had read extolling Hong Kong's riches? First she was crowded into the Huang's home, now she had to settle for a tiny apartment clearly needing a lot of cleaning and a lot of paint. But it gave them space and privacy. The monthly rent was frighteningly astronomical compared with the sixty *fen* they paid in China, but Mr. Liau assured them there was nothing cheaper anywhere in Hong Kong. They signed the lease with apprehensive gratitude, only years later discovering Mr. Liau had stood financial guarantor for them, without which they would have had no home.

"By the way, I know someone who has a painting studio. Would you be interested in that type of work?" asked Mr. Liau, visiting them two days later.

"I am willing to do anything," replied Anton, clearly interested. "I was commissioned to do several paintings of Mao Zedong in China," he added proudly.

"Really?" said Mr. Liau, raising his eyebrows quizzically. "Well, skill would be useful, but I don't think essential. Tourists are crazy about buying paintings here. I'll introduce you to the Ocean Painting Studio in Nathan Road, Kowloon. You just have to copy a few favorites pieces. It's easy work. You'll have to travel a bit, but the cheap rent in Yuen Long will make up for that. You can probably negotiate your hours with the boss. He might even let you work from home."

"This is better than I expected!" declared Anton gratefully, meanwhile thinking that "work from home" might allow him opportunity for musical work.

"Anton is an excellent artist," said YuXiang proudly. "He was in demand in China."

There was an awkward pause. "People here aren't too keen on portraits of Mao," Mr. Liau observed stiffly. "Anyway, painting should give you work to start with. Hong Kong is a bit fussy about accepting Chinese qualifications," he added, thoughtfully rubbing his chin, "but you could try visiting various schools to see if they want a physics teacher. I'd give the Baptist College a try first, and see how you go from there. I'm sure you'll get something soon, but this painting work is pretty easy to start with."

Clearly Mr. Liau did not regard painting pictures as a serious long-term occupation, but Anton enjoyed painting and thought this a great opportunity, one that could lead to an outstanding future.

Next day, after a long bus trip from Yuen Long to Kowloon, Anton had no trouble getting hired as a professional artist. He felt a surge of optimism, and decided his fears for their future in Hong Kong were unfounded. With his usual dedication he set about his assignments painting full-sailed junk boats (now very rare in Hong Kong, but tourists loved them); streets scenes of Hong Kong Island (which Anton had yet to see); and classic portraits of elderly Chinese peasants wearing wispy goatees and conical straw hats that reminded him of his long bus trips to the home of his parents-in-law. He enjoyed painting the portraits, found the street scenes a bit tedious, and wondered why people were still interested in ancient junk boats. After three days of feverishly painting (he was paid by the piece) he decided on the strength of assured work and a settled home it was time to achieve the reason for the move to Hong Kong.

"I will go to the immigration office tomorrow," he announced to YuXiang, when he arrived home that evening. He dared not tell her how angry his employer had been when he asked for the day off. What had happened to the "work at home" hours that Liau RiSheng had mentioned?

And now, here he was on his way to the immigration office. After his two-hour Lo Wu experience, he was anxious.

The journey reminded him just how far Yuen Long was from the center of Hong Kong life. A thin grey dawn was just breaking as they rattled through the fishing village of Tuen Mun (rumored to be the next New Town dormitory suburb of Hong Kong) and along the attractive shoreline towards Tsuen Wan. As they neared the bus terminus beside the Tsuen Wan Ferry terminus he could see the grey outline of Tsing Yi Island looming across the pale silver water of the harbor. He looked at his watch. It had taken one and a half hours to get to the Ferry terminal. It would be at least another hour before he got to the Hong Kong Immigration department. Fortunately, although he shivered with cold on the ferry trip across the harbor, the journey was smooth.

Mr. Huang's map was clear and accurate, and Anton found the Immigration Office without trouble. But although it was an hour before the office opened, already a long queue of people stretched ahead of him. He had hoped to be first in the queue, to get back to his painting work. But he had caught the first bus from Yuen Long, so there was nothing to do but wait. When the doors of the immigration office eventually opened, he was given ticket number thirty-four.

"That's a really unlucky number," said his queue neighbor with malicious glee. "You won't get anywhere with that! At least I have a lucky double! Any number with four in it means death!" he cleared his throat and spat on the pavement.

Anton ignored him, and found a place on the hard benches inside the office.

There he waited. And waited.

Children played chasing around the benches provided for clients and tripped over his feet. He tucked these unfortunate appendages under the seat, but it made little difference. An old woman had a high-pitched conversation with the deaf old man beside her. Eventually the old man shuffled out and returned with some powerfully smelly fish balls in a plastic bag. The old woman ate hungrily, but an official banged on his counter and angrily shouted that she take her stinking food outside. She slid it under her seat, and whenever he turned his back, took another ball. A young man on the bench opposite spoke to her, and she passed her fish balls across to him. Hungrily he pulled out a skewer of four balls, and began chomping. The container went back under the seat, and the smell was thick and palpable. Two young men stretched out on the floor and dozed off. The official shouted at them too, but they slept on peacefully. Finally, another officer, who would have been better occupied with his job, came and angrily shook them awake. They sat up groggily. After much official yelling they got up and languidly draped themselves over a bench.

Anton was so engrossed with the antics of all these people needing immigration attention, that he missed the call of his number. Only when he heard number thirty-five did he realize his mistake. Hastily he got up and ran to present himself.

"You missed your call! Go home! Come back tomorrow!" At least the man spoke Mandarin. Anton feared they would speak Cantonese, which was unintelligible to him.

"But sir, I come from Yuen Long. It is a far journey."

"That's your problem!" snapped the official, and picked up the microphone to call another person.

"But sir, please! I need help!"

Anton's demeanor was so gentle and polite the official looked up. "Well, what number are you? Thirty-four? So, you just missed?"

Anton nodded vigorously, hardly daring to breathe. "Well, what do you want?"

Anton had prepared well. All his papers were in order. He pushed these towards the official, who glanced at them.

"I would like to apply for a Hong Kong passport, sir," Anton said deferentially. "My mother in Indonesia is very sick, sir."

"A what!" shouted the official, riveted to full attention. "What did you say? A passport? Did you say you want a Hong Kong passport?" He grabbed Anton's papers and began sifting through them angrily.

"When did you arrive in Hong Kong?" he demanded.

"Two weeks and four days ago, sir."

"What!" The smartly uniformed man put his head back and roared with laughter, long rolling guffaws of callous mirth. "Are you crazy?"

"Hey, hear this!" he called to the official in the next cubicle, and then referring to Anton by a highly insulting name, exclaimed, "This Ah Chan wants a passport! Can you believe it! He's been here less than three weeks!"

"My mother is very sick," repeated Anton. "I came to Hong Kong so I could visit her. It is on my official documents as the reason for coming here."

The second officer came to Anton's cubicle, and stared belligerently at him. While the first was still chortling, this man scowled. Anton found two officials crowded into one small cubicle very intimidating.

"My mother . . ." he began again.

"Stop! You've said that! We heard! You're all the same, you . . . you . . . Ah Suk! Useless mainlander!" bureaucrat number two snarled, using another insulting name. "You're all so ignorant. Don't you know anything?"

"But, sir, my mother is sick, she is dying," repeated Anton. "I am the eldest son. That is why I was given permission to come here!"

"Yeah, yeah, yeah! We've heard all that before from the likes of you. Every Ah Suk has a mother, and they're always sick. Ah Chan mothers are never healthy." He eyed Anton's well-worn clothes. "Now, let me tell you, there are no passports for the likes of you until you have been here for at least seven years, and behaved yourself to boot! Did you hear? Do you understand? S-e-v-e-n years!"

Seven years stretched before Anton like a black eternity. "But my mother! She's sick! She is very sick!" pled Anton, his voice squeaking with emotion.

"Look, those papers of yours suggest to me that you are very likely a communist agent. I know you were held up for questioning at the border. Yeah! See, I know everything! We don't like your types. A physicist with construction tools! Ha! Ha! Get moving! You're lucky they let you in! If I'd been out there you would have been sent back where you came from, quick smart! Come back in seven years, if you still want to visit your dear, poor, sick mother in Indonesia. Get!"

"But sir!" pled Anton. "Please, have a heart for my mother!"

The first official leered at him, a cruel disparaging smirk. "Now there is another way people like you can get a passport," he declared in an oily, conspiratorial voice. "If you have enough money you could buy a South American one, if you really want to visit your dear, poor, sick, mother! One hundred thousand dollars might do the trick! Ah Suks like you have heaps of money, and South American passports are easy to buy." Both officials turned away, their shoulders heaving with mirth.

"Look at his clothes!" chortled the first officer. "He'll be struggling to eat. He probably can't count past fifty."

"I've no idea why they let these dumb uneducated people in," said the second.

Anton stood, paralyzed by inexpressible anger, rooted to the ground with horror. Could he believe what had happened, what he had just heard? The immigration officers remained with their smart navy uniformed backs to him. Finally, they stopped laughing, walked away, and began discussing the next horse race meeting at Happy Valley.

Slowly Anton gathered his papers and returned them to his pink plastic bag. His shoulders drooped, his anger dissipated, and he walked across to the bench where he had sat all morning. Flopping down heavily, he closed his eyes, utterly defeated. When finally he opened his eyes and looked at the clock on the wall, it was eleven minutes past twelve. No wonder his stomach was rumbling; he had last eaten seven hours ago. But it would be another three hours at least before he could get to Yuen Long. He was far too dispirited to report for work at the art studio.

As soon as he walked in the door, YuXiang knew his mission had been a dreadful failure. Wordlessly she heated the kettle for a cup of tea.

As Anton sat in his chair sipping it, she silently continued re-winding wool from NeeZi's only cardigan, ready for reknitting.

Neither of them said a word. Only YuXiang's hacking cough broke the silence.

"Coming here hasn't helped her!" thought Anton angrily.

That afternoon Anton wrote a letter suggesting his parents visit Hong Kong. A month later his brother wrote saying his mother was far too ill to travel.

*****

Six weeks later Anton stood in front of his Ocean Painting Studio boss, trembling with anger. It was hard to make make Anton angry, but his boss had passed his limit. Anton understood he had not been employed long enough to get paid with other employees at the end of January, and must wait till the end of February. He coped with the long daily commute and the constant rudeness because he could not speak Cantonese. He managed the worry of his dwindling financial resources. But this was just too much.

"But sir," he said, holding out a pitiful handful of red Hong Kong dollars, "this . . .! This cannot be right!"

"It is the standard rate," the man said coldly, turning away. "Thirty dollars[1] a picture. Take it or leave it."

"But sir, you said I would earn enough to live! You sell my paintings for at least five times that. My work is good. I know you get more for my paintings. People all say my work is good. I was a commissioned artist in China. You promised I would earn enough to feed my family!"

"You're slow," the man sneered. "Why do such good paintings? Your work might be good quality, but we just want stuff to sell dumb tourists. All they want is a bargain, so why make masterpieces? Why kid yourself you're some kind of real artist?" The man cleared his throat noisily, and expertly spat through the shop door to the pavement, narrowly missing an elegant pair of European legs. "So you think you're a genuine artist, eh? They even let you paint Mao Zedong did they? Ha! Ha! Ha! That's all they paint over there! Better than us, eh? Well, let me tell you, we don't need Ah Chan artists telling us how good they are! Just someone churning out stuff stupid tourists buy! None of them knows what a good Chinese painting looks like!"

"But, sir . . ." began Anton.

---

1. At the time about US$3.

"We told you. Ah Chow told you several times not to use much paint, simply do something that looks good. I told you clearly you were paid by the piece. What I sell those pieces for has nothing to do with you!"

"But in China my work . . ."

"Cut it!" the man commanded, with a snarl. "I couldn't care less that some stinking flea-bitten Communist cadre thought you could paint! You either take that money and clear out, or I'll grab it back and you can go without! If you want to work here, it's my way, or starve. And remember, there aren't many places ignorant Ah Chans like you get can work around here!"

Anton closed his hand. The crisp new banknotes cut cruelly into his palm. Hastily, as though they were poison, he stuffed them in his pocket. Then turning, he picked up his case of paints and brushes from his spattered workbench.

"Goodbye sir," he said. "Goodbye."

"I'll see you tomorrow. I hope you have some sense by then!"

"No, you won't," said Anton decisively. "I have principles. I might come from China, and I might be poor, but I know my worth!"

The workshop owner threw back his head and laughed maliciously. "Just wait till you're starving. That usually talks sense into Ah Chans, and quickly gets rid of their stupid principles! Nothing like hunger to let you know what you are worth!"

Anton walked out of the shop.

He had ample time to think on the long bus ride home. In the six weeks of his "employment" he had earned just enough to pay for his bus fares, and his lunch. There was absolutely nothing over to pay rent, nothing for family food. A quick calculation revealed the meagre supply of money brought from China, with very careful budgeting, might last four months. No, he would not work for that man! He would never allow himself to do the shoddy work other painters did, others paid the same thirty dollars a picture he was. Suddenly he remembered he even supplied his own paints, so his wages did not cover even his bus fares and lunch!

I will do anything, but I will not work for that man! he said again to himself. As the bus pulled into Tsuen Wan, he made a quick decision. He remembered his thoughts on the train from Beijing. Music was where his future lay. He got off and went to his friend Huang.

"I'm willing to do anything, anything, just to eat" he said, as they sat sipping tea. "But this painting does not even pay bus fares and lunches. It's crazy work! It's wicked, like Dracula! Can you suggest anything, or ask Liau RiSheng if he knows any other work? I desperately need money for rent and food."

"It's not easy, my friend," responded Mr. Huang, thoughtfully stroking his chin. "People in Hong Kong think anyone from China is backward and stupid."

"But . . . But!" spluttered Anton, unclenching his fists, and wringing his hands. "Surely they know China has made tremendous progress, that we . . ."

"Yes, yes, yes, we know, we know, but they don't," answered his friend. "As soon as we open our mouths we don't belong. Even when we write, they think we are primitive and backward. Chairman Mao simplified the characters which meant almost everyone in China has learnt to read. But the Cantonese think we are too stupid to learn their complicated ancient-style calligraphy. What we think is progress, they think is stupidity."

"Well, I don't care how stupid they think I am, I won't do shoddy work. And, I won't work for nothing. I must have real work to feed my family."

"I'm not sure if you would like this," began Mr. Huang. "My aunt has a factory in Kwai Hing and if I spoke to her I could get you work there."

There was a long silence. Anton sat quietly, hands folded in his lap, eyes fixed on the floor. Finally, he looked up.

"You are very generous," he said. "I am very grateful. But as we have discussed I know I must try to do what I know I can do. If I start in the factory, I may never get out. I . . . I have talked to others. I know what happens. No . . . I need to . . . I want to . . . I believe I should pursue the path I know. If I try hard enough I can make something work. I have this feeling something in music is my destiny. I was a scientist, and loved it. Now my path will be with music."

There was another long silence. "I admire your courage," said Mr. Huang. "It will be hard for you, really hard, especially at first."

Anton laughed bitterly. "Harder than it's been already?"

"Yes."

"Maybe I am just arrogant. But let me try, let me try music, at least for a few more months."

"I understand. I'll talk with Liau RiSheng. You remember he's a musician? I'll do everything I can to help you."

"Thank you," said Anton, as he stood to leave. "Thank you very, very much."

"By the way," said Mr. Huang, as he opened the door, "how long were you planning to come to Hong Kong before you actually got here?"

"I wasn't really planning to come. I loved my work in China. You know I was a respected research physicist there. I was beginning to establish my reputation as an acoustics physicist. But my mother in Indonesia got sick.

So to answer your question, it was three months between my application to leave and the time we arrived here."

"Yes, I thought so. I was planning for more than five years, and saving all that time. My auntie here in Hong Kong warned me how difficult it would be."

"But I have no one here in Hong Kong, no family at all; just a very sick mother in Indonesia."

"I know. I know. I'll try to help you."

*****

But Mr. Liau had no new ideas and Mr. Huang's next suggestion proved no more successful than the painting fiasco. He and Anton decided that since Anton was a talented wood worker he should apply for short term carpentering positions. Anton had no trouble being accepted for work, but when he arrived with his box of hand tools he received guffaws of laughter from other workers.

"Of course, mate," two carpenters said to each other at his first job, as they plugged in their electric drills and planes. "They don't have electricity in China."

After they stopped laughing at their own joke, one said, "They eat their own filth up there you know. They're like animals." This caused more peals of mirth. Meanwhile Anton stood silently waiting to learn what he was expected to do.

"Thank you for showing us your prehistoric equipment," the foreman said with exaggerated slowness, amidst snickers from other workers. "But we aren't cavemen here. We use tools. Take your stuff back over the border where it belongs, but get out of here with it."

It was the same everywhere he tried to get carpentering work. He applied for more than a dozen positions but without success. Sometimes he managed to get a few days' work repairing broken cupboards or other work no one else wanted. He even gained a small reputation for skill in difficult cabinet making. It helped his meagre savings last a little longer, but did not solve the problem of providing a real income for his family. In China they had little, but they had enough. In Hong Kong they did not even have enough.

The Sie family plight, typical of many other immigrants from Mainland China, was a source of merriment to many Hong Kong people. It was the topic of a comedy movie immensely enjoyed by many local Hong Kong residents who believed their contact with things British somehow made them superior, although the film was severely criticized overseas. A Singaporean critic called it a "shameful movie for all Chinese." The movie told the story of a poor worker, A-Chan, who fled from Guangzhou to find work in Hong Kong. When his employer asked how he was coping, the man admitted he had not eaten for two days. A plate loaded with one dozen McDonald hamburgers was produced, and his fellow workers incited him to eat more and more of this strange western food. Anxiously the hungry man ate all twelve burgers, but his eyes popped from their sockets and his throat was clogged with food, causing taunts and merriment from fellow workers. For Anton, this film was a source of shame and anguish.

Anton was not paid enough to live by artwork. He was not allowed to be a hand-tooled carpenter. The few small handyman type jobs he obtained in the Yuen Long were not enough for the family to live. And despite his belief that his future lay with music, he had found no musical position. Things were getting desperate. Like A-Chan in the film, he and his family faced starvation.

One day when he arrived home, tired and dusty with nothing to show for days of tramping between prospective employments, YuXiang met him at the door.

"Can you give me ten dollars? YanYan's school . . ."

Anton walked straight past her, then turned. "Woman!" he snarled. "I've already given you seventy dollars this week! Leave me alone! I cannot give you one *feng*! My life is ruined! I wish, how I wish, oh how I wish with all my heart, I had never left China!"

YuXiang had never heard her husband speak like this. He was always so focused, so positive. He sank on a box, clasped his head in his hands and shook.

Hong Kong had been hard for her too, really hard. Daily she was confronted with an abundance of food she never dreamed existed when she was in China. But whilst in China they had little, now they had nothing. The seventy dollars Anton gave her each week to feed them went nowhere. Worse, she became aware that market hawkers regularly cheated her. She learned to watch their balances very carefully, to let them know she was watching, so she got fair deals for her money. She spent much time walking around the market looking for the cheapest fruits and vegetables. She was creative: fruit other buyers rejected she bought and used, like freezing bruised plums (in the ancient refrigerator the Huangs had given them) to make delicious natural ices. But she could only long for the beautiful clothes in the markets, and for the foods so freely available to anyone with money. And much more seriously, her health troubles had not improved. Although she tried to hide the blood stained mucus she daily coughed, she could not conceal her loss of weight. Anton had always been solicitous of her health, but now he never mentioned it. She thought Hong Kong would be the beginning of a new life, not this!

Waves of misery washed over her as she looked at her husband. Then suddenly she thought, "But he carries the responsibility. It must be terrible for him." She knew what she must do.

"The girls are good students," she said. "I think Hong Kong will give them opportunities. And I have faith in you."

Anton looked up. YuXiang smiled at him.

"Husband, I believe in you. I agree that you should try to find your place in the music world of this great city. I did not complain when you decided not to take work in Huang's aunt's factory. But music is not so easy to begin with. I wish I could work. I have been looking..."

Anton looked up, very surprised. He had no idea that his wife had been trudging Yuen Long streets looking for work.

"Yes, I have been looking. I will keep on looking, but right now there is nothing for me."

Anton put his hand in his pocket. He pulled out ten dollars and thrust it towards her.

YuXiang smiled. "No," she said. "YanYan doesn't need to go on the school trip. I will write a letter."

Anton put the note back in his pocket.

*****

Anton responded to the gentle knock, and, cautiously opening the door of his Yuen Long apartment, saw Liau RiSheng and his wife Chung Ping.

He greeted them hospitably, and offered the only two real stools the family possessed. Searching rubbish collection depots around Yuen Long Anton had found some wooden boxes and old discarded cushions. YuXiang carefully unpicked the cushions, and washed the covers and wadding as she had done for their clothing in Changchun every year. Then from the old cushions she made seats for the boxes Anton found. Between them they made serviceable stools from other peoples' junk. Anton kept searching rubbish depots until he found a pair of scratched but otherwise sturdy stools; these they offered guests.

"Come in! Come in! Welcome!" Anton cried, playfully using Indonesian.

Mr. and Mrs. Liau smiled, returned the Indonesian greeting, and sat down sedately on their stools, hands folded. YuXiang began boiling water, hoping desperately there were still tea leaves in the canister.

"Sie SinSan,[2] we are worried. You are looking very thin," said Mr. Liau.

"Yes, you have lost weight," added Chung Ping. "I know you haven't had success with work here in Hong Kong. I've been thinking about you a great deal. I think you can get music work."

"How?" Anton asked wearily. "I have walked everywhere. I tried every music center you told me about. I offered to teach, I offered to repair instruments."

"Who have you contacted?"

Anton sighed. "The Baptist College."

"Yes, yes, but didn't you try to get work there teaching physics?" butted in Mr. Liau.

"Yes, I tried everything! They absolutely refused to accept my Chinese certificates. They behaved as though they were forgeries. We want real qualifications, they told me."

"But I'm talking about music," persisted Chung Ping.

"I know, I know," said Anton, slightly irritably. "When they wouldn't listen to my offer of teaching physics and math, I told them I could teach violin."

"And?"

"Well, they didn't quite laugh. They just said 'What certificates do you have for that?' and it was the same old thing. No English certificate, no work."

"Who did you see there?"

"Somebody in the office, a very young woman. She refused to let me talk to anyone else."

"Did you ask to speak to someone else?"

"Do you think I am stupid?" returned Anton impatiently. "Of course I did! She refused to give me an appointment to see anyone else."

"OK. Who else have you talked to?"

"Every music shop in Hong Kong. I first offer to teach violin, then I offer to repair instruments. It's always the same. Where is your certificate?"

"Every shop?"

"Well, I can show you the list, and if you know of any other place, you tell me." Anton got up, and took a sheet of paper from inside the Carl Jalovec volume that lay proudly amongst curls of wood shavings on the

---

2. *SinSan* is Cantonese for Mr.

small workbench he had made in a corner of the room. Pieces of a half made violin lay beside it. He handed the paper to Chung Ping.

Anton gestured towards the pieces of violin. "I thought perhaps I might be able to make and sell a violin," he sighed apologetically.

Chung Ping took the list and glanced through it. She shook her head sadly. "No, I do not know of any other place. But there must be more."

For several minutes the only sound was four people sipping extremely weak tea.

"It's really hard," Chung Ping finally said. "But perhaps I can help a little, just a little. Would you be willing to teach my students music theory? Most of them hate it, and I'm not good at teaching this. But if my students are to progress beyond grade five in their piano performance examinations, they must have a music theory paper."

"I'll try anything," said Anton firmly.

"How is your language coming on?" asked Mr. Liau.

"You mean Cantonese?" answered Anton. Mr. Liau nodded. "Well, it's not," said Anton. "I can't speak a word. But I can speak English. Maybe..."

"English!" exclaimed both Mr. Liau and his wife together. "Why didn't you tell us? Can you really speak English?"

"Yes, of course I can. But I thought you knew? How else do you think I could talk to the people at Baptist College?"

"This could change things," said Mr. Liau. "I wish I could speak English. It must surely open doors for you."

"Are you willing to go to the homes of students I find?" asked Chung Ping.

"Like I've said so many times, I'm willing to do anything," said Anton.

"Well, here is the address of one of my students. He's a good boy, a fifth form student so he should be able to speak some English. You can charge him two hundred dollars for ten lessons. He might need more."

"Two hundred dollars!" said Anton, quietly to himself. "One month's rent!"

Across the room YuXiang heard, lifted her head, and smiled.

"If you like, I can find more theory students for you," said Chung Ping.

"I would like, very, very, much," replied Anton.

*****

Anton might have been struggling to find paid work to support his family, but he was not idle. NeeZi and YanYan attended local schools as soon as they arrived in Hong Kong, YanYan in primary, and NeeZi at middle school. As in China, Anton took a strong interest in their violin-playing development. In his opinion, NeeZi was progressing well, giving promise of becoming a top quality violinist.

YanYan was a diligent student, but a little more independent in her attitudes. NeeZi was willing to ask for help, but YanYan preferred to puzzle things out on her own. Soon after she started school in Hong Kong she shocked Anton. A test question (she had many tests) was "from which direction does the sun come up every morning." From her Changchun experience YanYan knew that the sunny side of any place was south, so she confidently answered "south". Anton looked at her in disbelief. As his small daughter watched his frowning face she heard him mutter something that sounded like, "How wasteful for the sun to shine on you for the last seven years!" A crestfallen little girl determined she would learn better in the future. She discussed ideas with her father to learn from him, but made sure she did not need his help. And as for practicing her violin, all Anton needed to do to get her working was to fix his large expressive eyes on her, and she was scurrying to get it.

One day Anton read in a local paper found lying on the pavement that there would be a junior violin competition in a couple of months. The pieces chosen for the competition were well known to him, and entry was free. He was determined that NeeZi should enter the competition, and began coaching her for it.

With their lack of Cantonese, and even having to relearn their calligraphy, school was not easy or fun for the girls, despite their having been top students in Changchun. Learning to write again was very discouraging, especially as the calligraphy seemed so similar. NeeZi could see no sense in the complicated traditional Cantonese pictographs. But she happily retreated into the world of music, basking in the approval of her teacher-father. By now NeeZi had progressed so far that she enjoyed playing violin and needed no pressure to practice. She also realized her music brought joy to her father. In China he was always busy, so busy she hardly saw him. He was happy in China, and she knew he held a highly respected job. Now he was not busy, and he often looked stern and angry. He never said anything to her, but she had seen people treat him rudely in shops, even more rudely than her schoolmates treated her. Her parents did not speak openly about it, but she wondered if her father had any job at all.

So when Anton announced, "NeeZi, I want you to enter the junior violin competition in Hong Kong," she happily agreed. The joy that flooded his face was worth every bit of effort she put into her practice.

A week later Anton was sitting on one of the padded boxes listening to NeeZi's practice performance.

"That was good," he said, somewhat absentmindedly. "I'm sure you will do well. But you'll need an accompanist. I don't know anyone who can help with that." He sat, puckering his brow in concentration as he tried to think of a solution to the problem.

"What about that lady? The one who gave you the music theory student," suggested NeeZi.

"No, it has to be a young person," replied Anton.

"But Father, didn't you say she's a piano teacher? She must have at least one good student that could help us."

Anton looked at his daughter in surprise. "Why, you're right, of course. I'll go and visit her right now."

"Now? But won't she be teaching now?"

"Yes, but that's good. I'll listen to the students myself, and see if there is one good enough for you."

NeeZi smiled. Her father rarely praised her. Then he would say little things like "good enough for you," that assured her he was very pleased with her.

"I'll practice well while you are gone," she smiled.

When Anton arrived at Chung Ping's house he could hear the piano. The frequent mistakes and murmur of a female voice made him decide this pupil was certainly not a suitable accompanist. Finally, at his insistent knocking the door opened, and a slightly annoyed Chung Ping frowned at him. Rapidly he explained his errand. She relaxed, smiled and invited him in.

"I have two students I think would be suitable, and you are very fortunate. Both have lessons this afternoon."

Anton settled on a chair in the corner of the room. He had brought a newspaper, salvaged from the footpath, but decided its rustling would distract student and teacher, so he just quietly listened. He was impressed with his friend's teaching ability. Although her Cantonese was poor, she had learned enough to clearly direct her students. He wished he could find some students for himself. But it was impossible. Without British qualifications, there was no chance of teaching for him. But perhaps, perhaps, someday he might.

"The next two students are the ones I recommend," announced Chun Ping, shaking him out of his reverie.

"Oh! Oh, good." Anton sat up straight.

The first was a teenage boy. He played with conviction and skill. Anton was sure he would do a good job for his daughter. Then a teenage girl arrived. She was early, but apparently as pre-arranged let herself in to her teacher's home and sat quietly observing the other student. Anton remembered the boy had been a little late.

When both students left, Chung Ping turned to Anton. "Well, who do you think you would like to accompany your daughter?"

"They're both fine young pianists," he replied. "The girl: does she always arrive early?"

"Always. I notice she listens carefully to what I tell SingNam.[3] She's eager to learn. He believes he is a good student and does not seem to mind her being there."

"I think the young girl would be best."

"You have made a wise choice," smiled Chung Ping. "Can you let me have the music so I can go over it with LanLan,[4] and see what she says. I will visit her tomorrow, and let you know if she agrees."

"Thank you," answered Anton, picking up his crumpled newspaper, and opening the door to go.

LanLan, a little reluctantly, agreed to be NeeZi's accompanist. The first practice together went well, and best of all, the girls became friends and enjoyed playing together.

Finally, the great day of the competition arrived. Anton polished NeeZi's violin till it shone. She had long outgrown the quarter-size violin that brought Anton to the attention of the Beijing Music School. Now she used one of his handmade full-size violins. Anton checked every string, made sure there were replacement strings, and there were two bows in her violin case.

YuXiang had also done her part. At a traditional-style bed quilt factory in a back street of Yuen Long she discovered offcuts of pale peach-colored silk. With much gesticulation she convinced the factory proprietor to let her have this pile of useless (to him) strips of fabric for five dollars. Carefully she hand-stitched the strips together to make an interesting striped fabric, then from this pieced material made a simple slip dress. In the local dry-goods market she found some cheap but nicely patterned scarves, which, with careful cutting, made an attractive blouse to wear over the top of the silk dress. NeeZi looked stunning.

The whole family could not afford the bus fares to the concert hall, so YuXiang and YanYan were left at home. NeeZi sat nervously beside the window, while her father enjoyed pointing out city sights she had never seen before. At their destination was a crowd of young competitors, and a seething mass of parents and supporters.

---

3. A pseudonym.
4. A pseudonym.

As Anton sat in the hall listening to the performers, he knew his daughter would have no easy task. But he had taught her to focus on the music, not the audience. Her job, he told her, was to bring the audience into the music, to make them feel part of the mood and ideas of the composer.

At last NeeZi's name was called, and she stepped on to the stage with LanLan. Expertly she tuned her violin, and within a few bars Anton knew his daughter was at her best. No matter what the judges might say, she was the winner in his mind.

Generally a patient man, Anton found waiting for the the judges' decision very difficult. He looked around the vast throng but could recognize no one. Not that he expected to find friends, for Chung Ping had told him she could not come till later in the afternoon.

At last a hush descended on the crowd of excited teenagers and their supporters. The line of judges marched solemnly on the stage. The spokesperson stepped to the microphone. After an interminable speech about the quality of all performers and the beauty of music and the difficulty of deciding what criteria to use and the importance of unanimity and so on and so forth, ad infinitum, the man finally stopped and took a deep breath.

"And now," he said, "I have great pleasure in announcing the winner. The winner of the prestigious Junior Violin Competition of Hong Kong is Sie NeeZi."

There was a brief silence, then polite clapping. With a spasm of pain Anton realized no one knew his daughter, and all the favorite performers and their families were wondering who this unknown Sie NeeZi was. But as she came on to the stage to receive her prize, her classic beauty was striking and the audience broke into thunderous applause.

Anton could not contain himself. He jumped up, and with feet barely touching the ground hurried up the long aisle, to backstage. He must be the first to congratulate his daughter. His faith in her had not been misplaced. For the first time since he had been in Hong Kong he wished he did not look quite so ragged. But NeeZi would know he was wearing his best shirt and trousers.

To his horror, when he got to back stage his progress was blocked by a pack of paparazzi busily snapping pictures of NeeZi as she emerged.

"My daughter!" he cried loudly in English. To his surprise the photographers and reporters made way for him, letting him stand proudly beside NeeZi.

Next day newspapers across the British colony, both Chinese and English, blazoned the story of the "little girl from the mainland" who had won first prize in the Junior Violin Competition. The story was even broadcast on Hong Kong television news, a proudly beaming Anton beside his shy but beautiful daughter clutching her highly polished violin. In 1977 few people, it seemed, believed any good thing could come out of China. Now suddenly here was proof that it could. NeeZi's win created quite a sensation. Something good *had* come out of China!

*****

"It's incredible!" exclaimed Chung Ping perched on a stool sipping insipid pale lemon-colored tea with Anton and YuXiang. "It is just incredible! Do you know in the last two days I've had four more requests that you teach music theory to my students, and when I went to register some students for exams last week the office woman asked if I knew who taught 'that mainland girl' and where you were teaching violin! I told them I would find out. Apparently they've had several requests from prospective students wanting to learn from the teacher of the girl who won the violin competition!"

Scarcely believing what he heard, Anton sat motionless.

"It would be good to have more theory students," he finally responded. "I think you have given me seven new names, right? But how can I teach violin if no one will let me? I still don't have any qualifications."

"Would you teach students in their homes?"

"Of course, but what difference would that make?"

"Well, the music schools might need qualified teachers, but if you go to the homes of students, I think it might be different. So, can I tell the exam office that you only teach privately, but if they give the students' names to me I will pass them on to you?"

"Is that true? Of course I'll go to the students' homes!" Anton raised his eyebrows incredulously.

Chung Ping rose to leave. "Your daughter's success has been good for me too, you know!"

"Really?"

"Yes, comments about NeeZi's good accompanist have gained me several more students."

"It is good to help each other," smiled Anton.

The piano teacher opened the door and left just as NeeZi arrived from school.

"I heard what she said," she giggled as she came inside. "She was so close to the door I couldn't help it! I'm happy to help her, but Father, have I been a help to you? That's what I *really* want to know."

"Daughter, you have helped me, your mother, and our family, far more than you will ever know!"

NeeZi smiled happily. "Perhaps all those years of practice were worth it after all!" she laughed.

"You need to get on with your homework!" said YuXiang in mock severity.

"Yes, Mother."

"Anton," said YuXiang, after their daughter walked away, "I think we will have enough to live on now, after all, don't you think? I mean, if we are very careful?"

Anton nodded, slowly.

"Do you think you might even have enough to buy a sewing machine?"

Her husband grinned.

Suddenly Anton stood up. "Why not!" he almost shouted, to no one in particular. "We can do it, and we will! NeeZi! YanYan! Come here at once! We're going to celebrate!"

Two heads, wide-eyed with astonishment, simultaneously appeared around the door. YuXiang frowned, not in disapproval, but in wonderment. Celebrate? What was that?

"Don't all stand there looking so silly! Get ready! We will celebrate what NeeZi has done!"

"Oh!" sighed YuXiang.

"How?" said both girls at once.

"We will eat out, that's how!" declared Anton.

"Are you sure?" said YuXiang frowning uncertainly. "It's very expensive!"

The girls disappeared into the room, and came out combing their hair. YuXiang seemed glued to her padded box.

"Are you sure?" she repeated. "It's not something you usually like to do."

"Yes, yes! Get ready, please."

YuXiang ran a comb through her long, luxurious hair, tied it into a simple ponytail, and declared she was ready for celebration.

Solemnly the family marched down the main street of Yuen Long. When they reached a small, rather dingy-looking restaurant, Anton turned in.

"Here?" said his family in unison. "Is this where we celebrate?"

"Yes, there's a Mandarin-speaking waiter here, I've been told."

And there was. Ostentatiously the waiter wiped the grubby plastic tablecloth, and brought a menu. Only two other diners were seen in the gloom.

"I don't eat seafood," announced Anton. "So what do you have? This is a special occasion."

"No seafood? Really? No seafood?" said the waiter, shocked.

"No, never."

The waiter ran his fingers through his spikey hair, and then down the menu. He mumbled no, not that, no good, probably no good. Finally, he stopped.

"What about fried rice?" he suggested helpfully. "And sweet and sour pork?"

"Any chicken?" asked Anton hopefully.

"I'll try," said the waiter doubtfully. "I mean we have chicken, but with no seafood, I don't know."

Anton looked up. His three women had their heads down and were staring determinedly into their laps. "What's wrong with you all?" he asked, bewildered. "This is a celebration!"

YanYan suppressed a giggle. NeeZi made a strange snorting sound suggestive of an aborted laugh.

"Do you have barbeque pork?" said YuXiang, and raising her head with an air of desperation, she burst into giggles.

"Certainly!" said the waiter with relief. "Is that all?"

Anton nodded, and the man disappeared. Soon he returned with a steaming platter of fried rice. It smelt delicious, but also suspicious.

"You taste it, please," Anton beseeched YuXiang.

"It is lovely, delicious!" she said, smiling broadly. "But not for you. They have used oyster sauce." Anton frowned. "But don't worry, we will enjoy it. After all, it is a celebration! What a good idea you had!" To the waiter she said, "Please bring a bowl of plain rice."

Suddenly eight-year-old YanYan demanded, "Why does father always make such a fuss about food when we eat out?"

There was a stunned silence. "We haven't eaten out since China," began Anton.

"YanYan, that is very rude!" YuXiang spoke severely. "Father has a seafood allergy. We have to be careful so he does not get sick. It is always like this when we eat out. But come, let us all enjoy this special occasion!"

Just then the sweet and sour pork arrived, and the whole family began to eat hungrily. The waiter produced barbeque chicken ("Just like the train to Changchun!" declared Anton) and barbecue pork, which they enjoyed immensely. The waiter's experienced eye quickly summed the family. The girls were still in school uniform, ubiquitous Hong Kong white frocks: YanYan's with a small brown bow at the collar, and NeeZi's with a slightly larger red bow. Both parents were clearly not only recently arrived from the mainland, but struggling to find their feet in Hong Kong. He was both waiter and proprietor of this tenth rate little restaurant. He suddenly felt wildly magnanimous.

"On the house!" he announced triumphantly, placing a platter of steaming tofu and cabbage dumplings on the table. "And no seafood! Cook's special!"

"Thank you!" The family's mumble was sparse reward for his generosity, but their faces beaming with happiness were ample recompense.

"I haven't eaten like this . . ." began YuXiang.

"Since China!" finished Anton and they both laughed, as they walked home through the hot tropical evening. Bats flew recklessly around the streetlights, and an enormous, beautifully-patterned brown atlas moth the size of an outstretched man's hand settled on Anton's shoulder.

"Father," announced YanYan, "I didn't know you, I mean we, could ever be so happy!"

*****

"Three of my theory students took their exams today," announced Anton.

The family had finished their evening meal. The girls were spreading homework across the only table. YuXiang picked up her sewing and claimed one of the padded boxes. Anton stood beside his small workbench, deep in thought.

"Will they do well?" YuXiang asked, a little anxiously.

"Yes, very well."

"Your teaching has made a big difference," she smiled.

Anton nodded. It made him feel very good to see the smiles on his wife's face, and the new shoes she and his daughters were wearing. He was still surprised that YuXiang bought shoes with the first clothing money he was able to give her. He liked the cloth shoes her mother made for them. But despite his acupuncture treatments, she still coughed badly every day, especially in the mornings, and she still did not have a sewing machine. That was one thing he really wanted to buy her.

"Yes, it has made a difference, but I need more students. Most students do not like music theory, and don't continue with it. It's not very stable employment, not like teaching violin performance. I have three violin performance students here in Yuen Long, but it's not enough. I must get more regular work."

"Yes? How? Haven't you tried all the schools?"

"I'm going to check out that small music school in Tsuen Wan, the one Chung Ping told me about. But I really want to get into violin repair work. I think that is where my talent and interest lies. I have an idea. But it means I'll be away till very late at least one night a week. I plan to contact the orchestra players on Hong Kong side, and try to get work from them."

YuXiang shook her head. "That's a long, long journey for you. Do you really think there's repair work available?"

"I've been thinking about it for months. I think this will be good for us all. I'll try again to get repair work from the music shops."

"But . . .!"

"Yes, I know, I know. No one wanted me before, but now I can tell them about NeeZi."

So every morning, before he began his afternoon teaching circuit, Anton trudged around music shops in Kowloon, Hong Kong and everywhere else. Day after day he was rudely dismissed by the mantra he had heard so often: we only use *qualified* staff here, thank you! He would arrive home tired and sweaty, with just enough time to wash and change before beginning his round of teaching his precious three violin students and the remaining music theory pupils.

One day Anton visited another music shop in Kowloon. The brusque dismissal was very familiar, the same old story about qualified staff. He was qualified, Anton gave a rue smile, but not by the right people in the right country. Suddenly he remembered that nearby was the famed Hong Kong bird street. Why not, he thought. He was now earning enough for the family to live, just, and YanYan's comment after the celebration dinner still rang in his ears. He did know how to be happy! How dreadful that his daughter thought that he did not! But when he thought of happiness he could not help but remember the birds he had enjoyed as a child. A bird did not cost much to keep. Anyway, there was no harm in looking at them. He collided with a hurrying pedestrian as he abruptly turned around and walked towards bird street.

He could hear bird street long before he reached it. The cheerful sound of myriads of birdcalls drew him like a magnet. As he wandered along the narrow street the desire to have birdsong greet him in the morning grew stronger with every step.

Suddenly he saw the most beautiful bird, and knew this was the one he wanted. Its glossy black head and back gleamed iridescent blue in the sunlight, and its rufous breast was cheerfully exuberant. Hidden shyly under its tail were pristine white feathers. Best of all was its gloriously rich and melodious song. Yes, he would have this shama. Willingly handing over money perhaps better used to buy new shoes for himself, or even the long-overdue sewing machine for YuXiang, he carried his treasure home with joy.

"Things must be getting better," smiled YuXiang when he walked in the door, carrying his prize purchase. "Did you get some more work?"

"No, but this bird will be good for our family. I know."

"Hope so," responded YuXiang, unenthusiastically.

But Anton's persistence finally paid off. Mr. Lee Hoi, owner of the Sheng Hoi Piano Company, agreed that Anton could take six old, broken violins and try to repair them. He also gave Anton some general carpentry work his shop needed. Anton was not surprised to discover that Lee Hoi was mainland Chinese. Originally from Guangzhou, he had been in Hong Kong for many years. He played in the Hong Kong Chinese Orchestra, so he and Anton had much in common, and became firm and lasting friends.

The small music school in Tsuen Wan also gave Anton a chance. The owner of San Lam Music House was again originally from China, and willing to give his struggling compatriot an opportunity. The school was not large, prestigious, or conspicuous: simply a small office unit in an industrial building in Kwai Chung. The constant grumble of machinery from nearby factories was hardly an ideal setting for a music school. The students were all very young, and required a great deal of Anton's patience. One of them was a seven-year-old Filipino boy named Luiji Oliva, whose father was the head of the pathology laboratory at Tsuen Wan Adventist Hospital. Lui, however, proved an apt pupil, and showed considerable promise.

But the long, tedious, weekly trips to the back door of the orchestra concerts proved more than discouraging, despite Anton's persistent determination. He obtained some simple business cards for this venture, and as the string players emerged after their concerts he presented them with his card offering instrument repair. Elegantly attired in their evening costumes, and flush with the success of their performance, most would take the proffered cards and immediately disdainfully throw them away. One or two performers, seeing Anton bend over to carefully retrieve the discarded cards, even dropped them, stood on them, and ground them into the filth of the street, making it impossible for Anton to reuse them. Anger rose in Anton's heart, but he neither said or did anything at this abuse.

One night as a dispirited Anton turned to leave, to his surprise a violinist came over, extending his hand in welcome.

"You're from China?" this unexpected friend asked in English. "What are you offering?"

Anton gave him a card, which the violinist took, and explained that he had expertise in violin repair work (well, he had successfully repaired the six old violins from the Sheng Hoi Piano Company!).

"You have an interesting accent," the violinist commented. "Were you born in China?"

"No, I was born in Indonesia."

A broad smile split the violinist's face. "So was I! I'm Lim KekTin. Here's my card. Come and visit me some time. I'll tell you about my family."

Anton was speechless. He grasped Mr. Lim's hand and shook it warmly. "I'll come," he said, too shocked to say more. But Mr. Lim lived a very long way from Yuen Long, on the other side of Hong Kong Island, and it was months before Anton accepted his invitation.

One night, idly looking over a discarded program sheet he found in the hall foyer while waiting for a concert to finish, Anton noticed the concertmaster for an up-coming performance was a Mr. Jan Van den Berg. Suddenly his years growing up in Indonesia flooded back. This man is Dutch, I am sure his is Dutch, Anton thought. He conceived a wild and daring idea.

A month later, at the end of the Van den Berg concert, Anton was very focused. He wasted no time giving out business cards, but persistently asked to speak to Mr. Jan Van den Berg. The elegantly dressed performers as usual looked disdainfully at the shabbily dressed man, and ignored him. Lim KekTin seemed absent. But a young trombone player finally directed him to Mr. Van den Berg.

"*Gelukwens,* maestro!" saluted Anton.

The well-built European stopped and turned, surprised to hear his native Dutch. "Yes?" he said sharply, using English with a thick Dutch accent.

Anton repeated his greeting, and, continuing in English, gave his frequently rehearsed offer to repair violins. This time the proffered card was accepted. "Who are you? How did you learn Dutch?"

Struggling to conceal the joy acceptance of his card gave, Anton explained that as a child he went to a Dutch school in Indonesia, and, newly arrived in Hong Kong, was offering his services as an expert violin repairer. Van den Berg nodded kindly, and told Anton he would let him know if he ever needed any repairs.

The long train and bus journey home passed in a blur of joyous optimism. At last his card was not trampled in the mud! He scolded himself for not visiting kindly Mr. Lim, and determined to do so as soon as possible. A few days later he phoned his friend and made the long trip to the other side of Hong Kong Island to visit the Chinese Indonesian violinist.

"Come in! Come in!" said Lim KekTin when he opened the door to Anton's knock. He ushered Anton into his comfortable home, and led him to a soft chair.

"How long have you been in Hong Kong?" he asked in Indonesian.

Anton told his story, and why he was in Hong Kong. Mr. Lim frowned. "My brother is in exactly the same situation. Because he worked in China he can't visit our parents. You may have heard of him. He's been professor of violin in Beijing. His name is Lim KekTjiang."

"Yes! I know him!" exclaimed Anton. "Well, not personally, but I know who he is. He's a great violinist. You see, every few months I used to go to the Central Musical Instrument Research Centre in Beijing and assist with research there. Of course I've heard of your brother!"

"Well, fancy that! You know, I have two other brothers and we all studied at the Conservatorium in Amsterdam. Lim KekBeng is a cellist in Holland, and Lim KekHan is a violinist, and like KekTjiang, also conducts. Did you know that KekTjiang was director of the Hong Kong Philharmonic in 1974 and 1975? He made numerous violin recordings."

"Really? I wish I had known him personally."

"Oh, when he next comes to Hong Kong I shall organize a meeting. I don't need any repairs for my violin now, but if I do, I'll let you know."

Anton rose to go, his heart full of joy. "That would be good," he said.

Anton could not get home fast enough to tell YuXiang about meeting someone from Indonesia, someone who was a good violinist and had a famous brother. After this happy encounter he was encouraged to persevere in his efforts try to get to know Jan Van den Berg.

For months Anton faithfully visited back stage and asked to speak to Van den Berg. Always it was just a polite greeting, and then the maestro strode away. Sometimes Lim KekTin would stop to say a few words in Indonesian. But one evening Van den Berg had more time, and spoke to Anton for a few minutes. How long have you been working with violins? What do you know about them? Anton talked about his years of study in Changchun, his daughter's violin that drew the attention of the Beijing Central Music School, and his research into the harmonics of the violin. Van den Berg listened intently.

Finally, he said, "Come next week. I have some bows that need re-hairing. That is, if you are willing."

Desperately trying to hide his excitement, Anton solemnly indicated that re-hairing violin bows was just what he most loved to do. The following week Anton collected three bows for re-hairing. It was a simple task for him to mend the bows. Then he boldly decided to return the repaired bows during the concert interval, rather than the end of at the concert. He considered he had a good excuse to intrude on the sanctity of the orchestra's domain, and hoped he might meet other violinists. And he did. As Jan Van den Berg tested his new bow, other string players crowded around him.

"It's good," nodded the concertmaster, playing a simple scale. "Thanks for a good job."

Anton went home with two more bows to repair.

*****

"I've been thinking," Anton announced to YuXiang one hot summer's day when, dripping with sweat, he arrived home from Lam Tei.

Lam Tei was a small farmer's village situated between Yuen Long and Tuen Mun, home of the Keung family who loved music. Anton first went there to teach the older son music theory, and then one by one four members of this cheerful farming family and two of their friends began to take lessons in violin. It was a wonderful break-through for Anton. Mrs. Keung took lessons, all three Keung children, and their friends Miss Chen and Miss Siu. They were all struggling beginners, but far more important than the welcome money he earned from this family was the kindness and respect they gave him. After months of rejection and outright abuse, here was a family who made him feel like a human being, a valuable human being, not an idiotic A-Chan.

"You do plenty of thinking," smiled YuXiang, breaking the silence. "Thinking about what?" She waited patiently.

"I've been thinking," he repeated, and paused. "I've been thinking we should move to Hong Kong Island," he added in a rush.

"Hong Kong Island! How could we possibly afford to live there?" YuXiang was horrified.

"Well, I have six pupils at Lam Tei, and another six at the San Lam Music House in Tsuen Wan. Remember that Filipino boy I told you about? Lui Oliva? Well, his sister has started learning, and he introduced me to a friend of his, a European boy named Sven Östring. They are all promising musicians. Lui told me there are other children in the apartment complex where he lives who are interested in learning violin."

"But..."

"I enjoy teaching these children, but I can teach all of them in just two days each week. What I really want is violin repair and making work. To do that I need to be available to the orchestras."

"But haven't they all repeatedly rejected you? Over and over again?"

"Lately some of them have been more friendly, and as you know, some even asked me to repair their bows. Remember Lim KekTin? He's always friendly and I think he will introduce me to his brother. But Yuen Long is such a long way from Central. I think Hong Kong is where my work will prove to be, with orchestras, but it's very tiring going there."

"Can I have that sewing machine?" asked YuXiang suddenly, a mischievous twinkle in her eye. This was her litmus test for financial security.

Anton grinned. "Yes, you can. Just wait till we have a new place to live."

"Then maybe you can look at housing on Hong Kong side, but only after I have that machine. I have waited a very long time, almost eighteen years!"

"You're right," said Anton, with a faraway look in his eyes. "I didn't have much when we got married, did I? Nothing much has changed, has it?" He shook his head and grinned.

"But we've managed," said YuXiang kindly.

Within a week Anton found a small but suitable apartment in Tung Hing Mansion, on Queens Road West, Hong Kong. But the joy of this success was dashed when he discovered he could not rent it; nor, in fact, could he rent any apartment on Hong Kong Island. It was the same everywhere he went, with every agent he contacted: "No, sir, I'm sorry sir. You don't have sir . . ." What he lacked was a permanent Hong Kong identity card, and sufficient money in his bank account to satisfy prospective landlords that he could pay the rent. No amount of persuasive talking would convince the real estate managers that he would make a suitable tenant.

A few days later he visited a friend recently arrived from China, none other than Tseng XiangHo, the man who recognized the value of NeeZi's quarter-size violin in Changchun, and who was instrumental in introducing Anton to the Beijing Central Music Conservatory. Anton poured out his tale of rental frustration, and asked how Mr. Tseng had managed to rent an apartment as soon as he arrived in Hong Kong. What special trick had he used? Anton was still unaware that Mr. Liau had stood guarantor for him for the Yuen Long apartment.

As Anton poured out his Hong Kong tale of misery and rejection, Tseng XiangHo suddenly stopped him.

"But wait," he said, "talk no further! I can help you!"

"You? But how?"

"Don't ask me too many questions! Just know that I didn't arrive in China as a penniless student like you did. I came to Hong Kong with a bank account. I trust you, and I know you are a good man. You can use my account as guarantee for your rental agreement."

Anton was speechless. This was kindness he never expected. But as Tseng XiangHo continued to urge him, he knew it was right to accept this generous offer. After all, he would not be using his friend's money: just the security of his account to satisfy the agents.

So in 1982 the Sie family moved from their tiny Yuen Long apartment to another equally small one on Hong Kong Island. They indulged in the luxury of a taxi to transport themselves and their few belongings (including the precious new sewing machine) across to the island. But as their taxi came to the outskirts of Tsuen Wan they encountered another unexpected difficulty.

With much high-pitched discussion and gesticulation, and NeeZi's efforts at translation, they discovered green "country" taxis were not allowed to enter the territory of the red "city" taxis! Nothing could persuade the New Territories taxi driver to risk making the journey to Hong Kong Island. They had no alternative but to laboriously transfer themselves and their goods to a second taxi hurriedly hailed, and thus finally reach their new home.

With her daughters' help YuXiang found new schools for NeeZi and YanYan, and was pleased when they reported they liked these schools better. Anton's trips to orchestra performances meant he now no longer arrived home in the early hours of the next morning.

Yes, it was more convenient. But once each week, unless a number eight typhoon warning was in place (when everyone in Hong Kong stayed home), Anton made the long bus-ferry-bus trek to Lam Tei to teach the various members of the family that made him feel a human being. He was always home in time for his dinner, always praising the friendship of the Keung family. He gave up teaching at the San Lam Music House and made a weekly trip to the Tsuen Wan Adventist Hospital, where he taught an increasing number of students. Most learned violin, but he had one cello and one guitar student.

*****

One morning, soon after settling into their new home, Anton woke to the riotous singing of his shama.

"It's going to be a good day," remarked YuXiang. "That bird always knows!"

"It always sings beautifully," laughed Anton, "but things are getting better. They are not yet stable, but I feel a huge weight has fallen from me. We will survive."

YuXiang's hacking cough muffled her reply.

Anton's mornings usually began with work on the new violin he was making. But this morning he decided to check the mailbox first. It was probably too early, but he felt the urge to inspect it. He climbed the stairs chuckling softly.

"Look at this!" he said, pushing a very official-looking envelope with a British Queen Elizabeth stamp towards YuXiang.

"What's it about?" she asked.

"It's from the Royal Schools of Music in England. My theory students have done very well again, and because all my students have gained consistently high marks this letter invites me to apply for the position of examiner for the school!"

"That would be wonderful!"

"No! No! No!" Anton frowned. "Get another rejection? If I applied for this position they would want my qualifications. I even risk losing the chance to prepare students for exams. I don't have to give any qualifications to present students for examination. So I just write 'Dr. Sie' because that's what I was known as in China. Yes, they might not even accept my students if they knew I have none of their precious qualifications. I'll tell them I'm too busy."

"Wouldn't it be worth trying?"

"No, I've been through this too many times. British qualifications are all that count. It's British qualifications or nothing, and don't I know! You would think China didn't exist!"

YuXiang was silent. Suddenly the unexpected clamor of the telephone broke her reverie. Anton jumped to answer it.

"Of course, Mr. Van den Berg! . . . Yes, Mr. Van den Berg . . . You are welcome to come to my humble home . . . Any morning would be suitable . . . I teach in the afternoons. . . Yes, tomorrow would be fine . . . Yes, eleven would be most suitable." Anton returned the receiver to its cradle.

"He's not coming here, is he?" asked YuXiang, panic rising with the tone of her voice. "Not here!"

"I told you moving to Hong Kong Island was the right thing to do!" said Anton, grinning broadly and triumphantly ignoring her anxiety.

"What does he want?"

"He didn't say, except he wants me to check something. But he's never asked me to do this before, so he must have something important for me to do."

YuXiang cleaned and re-cleaned the little apartment. Anton's violin construction activity caused her constant discouragement. She was meticulously clean and tidy, but Anton's activities caused a constant supply of sawdust and wood shavings. She tried cleaning quietly after him, but he complained about the sound of her sweeping, mopping and dusting. This time he made no objection.

She stood back and surveyed her efforts. It was wonderful to see the house sparkling clean for once, and she decided the expense of a bouquet of flowers was warranted. Very early next morning she made the trip to the Queen Street markets. Then at 10:30am she picked up her knitting, went into the bedroom, and firmly shut the door behind her. Even if she listened at the door she would not understand anything, so she would just wait. At least she would be useful.

Mr. Van den Berg arrived at the Sie apartment punctually at eleven o'clock, cradling a case carrying a beautiful yellow violin. He wasted no time in small talk.

"This is a valuable violin, but it has never made me happy. Something is wrong. Can you check it?"

Anton took the proffered instrument and carried it to his workbench. He brushed aside a few wood shavings and sullied YuXiang's spotless floor. For several minutes there was no sound except taps on wood, and plucked strings. Finally, he straightened up.

"It's the base bar. It hasn't been inserted properly. I can fix it, but the violin will need to be opened up."

"I trust you. Do what you think best. I hope you can make it right." Jan Van den Berg stood up, dusting his immaculate grey sports trousers. "It's no good the way it is."

"I will do my best. Your trust honors me."

"I know you will."

As soon as his patron left, Anton began work on the violin. Carefully he opened it, and removed the troublesome bass bar. Over the next few days he made careful adjustments, and refitted the bar. By then Anton realized the importance of the work he was doing. This was a valuable Carlo Tononi instrument. Carlo Annibale Tononi was an Italian luthier who worked first with his father in Bologna, and then transferred to Venice. Bologna was the home of the composer Corelli, but early in the seventeenth century when Tononi moved there, Venice had become the more important musical center. Anton was aware that a genuine Tononi violin was worth tens, even hundreds, of thousands of dollars. This unexpected assignment allowed him to examine the workmanship of a great violin, and to test his theories about good violin construction. Finally, after two weeks' intense work and a great deal of anxiety lest he fail this important undertaking, Anton was satisfied the violin was performing in its original splendor. He found some new maple wood that exactly matched the original, and made perfect junctions on the top and back plates so that his repairs were virtually invisible. He returned the restored instrument to Jan Van den Berg during a concert intermission.

Anton had made the trip backstage many times with repaired bows. He quickly located the Dutch concertmaster clad in his impressive frilled white shirt and tie, with black tailed coat. Hurrying towards him was Anton, in his old open-necked shirt and worn black jacket.

"Please, sir, please try your violin now," said Anton, holding out the case containing the valuable instrument.

The maestro smiled, but said nothing. He took the proffered case, pulled out the violin, tightened the bow, and frowning slightly, played a few scales and chords.

"What do you think?" he said, passing the instrument to another violinist, apparently ignoring Anton.

"Jan, can you trust this Chinese immigrant? He would have no idea of the value of your violin!" a European violinist watching nearby sneered. There were murmurs of agreement from other members of the string section. Clearly they did not expect that Anton could understand them.

Anton stood there, suffering once again deep shame and pain.

Suddenly, angrily, Van den Berg turned on them all. "This man is an excellent violin repairer! Can't you hear how well my violin plays? It sings! At last, at last, it sings! He is much better than others. I spent large sums of money trying to get this violin right, but those famous repair shops did nothing! But Mr. Sie has done a very quick and sure repair. He certainly knows what he is doing! I admit I am surprised. I did not expect his work to be good, so very good. He's probably the best violin repairer I know! He'll get paid what others charge and more."

There was a stunned silence.

"Look at his work!" declared Van den Berg triumphantly. "Take it! Play it! Hear it! This man really knows what he is doing!"

Suddenly all the members of the string section of the Hong Kong Philharmonic orchestra were crowding around, begging Jan Van den Berg to let them try his restored violin.

"Do you know I took this violin to three of the top repairers in Europe," the concertmaster persisted, as though Anton was his treasured private property. "But none of them fixed this problem! Months and months they had my violin! And this man took less than two weeks to completely restore it! I am so pleased!"

"Thank you, sir," said Anton, gratefully.

"Do you speak English?" a performer asked in astonishment. "Where do you live? Can I have your card?"

That night Anton handed out every single one of his cards. Not one of them hit the ground!

Soon other violinists were asking for his help.

"Congratulations!" a gentle voice beside him said. It was Lim KekTin. "Jan is not easy to please. You must have done a good job!"

Anton could never thank Jan Van den berg enough for the support he gave.

*****

One day, soon after they moved into their apartment in Queens Road West, their friend Professor Chen MingJun from Guangdong Zhongshan University, with whom they had stayed on their journey to Hong Kong, came to visit for a few days. He was planning to accept an invitation from Hong Kong University to spend three months there as a research fellow. Both Anton and YuXiang were delighted to have him stay with them, despite the limited space in their tiny apartment. It encouraged Anton to learn that someone with Chinese qualifications was being recognized in Hong Kong. They also enjoyed finding and preparing Indonesian foods that were impossible for Professor Chen to obtain in Guangzhou.

"You should come and bring these delicious foods for my wife to try!" he joked.

But Anton shook his head. "I don't have a permanent resident card yet, and it might be difficult for me."

"Perhaps your girls could visit us!" laughed Dr. Chen.

"Could we?" responded YanYan eagerly.

Anton looked at her. Both girls had student ID cards because they attended Hong Kong schools. Perhaps it would be possible.

"I have my final exams coming up," said NeeZi hesitantly, not wishing to sound ungrateful, "and I need to practice my violin hard so I can be admitted into a conservatorium."

"But I don't!" declared YanYan triumphantly. "Please let me go! The summer holidays are so long and boring here! Mum's always sick and Dad's always busy!"

"I don't think so," a frowning YuXiang said crisply. "You're only ten and I need you at home to help."

But in the end YanYan won, and for several years every summer holidays she travelled alone by train to Guangzhou to visit her special aunt and uncle. She never complained about carrying heavy bags (sometimes weighing ten kilograms or more) of Indonesian food for Uncle Chen. Nor did she complain about the intensive math tuition she was given while on holiday. She suspected this was the bargaining point with her parents that allowed her to make the journey. Whereas NeeZi had excelled at math, carefully tutored by Anton in China, YanYan had many social distractions, and now Anton was always too busy to tutor her. But under Uncle Chen's patient tuition she blossomed, and soon her math was as good as her sister's. And of course, she went insect hunting with Auntie Wu. The variety of insects in Guangdong amazed her. Over a dirty puddle of water in the middle of some long grass she could find brilliantly colored dragonflies, some with red abdomens, others with blue, or green, or yellow. The blossoms of lantana vines attracted a myriad of bright butterflies, and poking around rocks and leaf debris was sure to locate several vividly colored beetles. But it was the aesthetic beauty of these creatures rather than their science that interested YanYan, and Auntie Wu's hopes that she might become a biologist were not realized.

When Professor Chen took up the research post with the Hong Kong University in 1983, he stayed with the Sie family, much to the delight of both girls. They were very happy to give up their beds and sleep on the floor in the living room for the three months he lived with them.

*****

Anton, however, was still often humiliated.

One day on the way home from his weekly trek to Lam Tei, he went to the Hua Guo Restaurant to buy a box of takeaway food. YuXiang's health was deteriorating, and he thought bringing home the evening meal would help her. Several restaurant employees were milling around the counter. Anton's Cantonese was very poor, but he had learned a few words, enough to manage in the market place. Three times he spoke to the workers, indicating he needed service.

"He! He! He! Ha! Ha! Ha! Listen how he tries to say please! Ignorant Ah Suk!"

"He's got his tongue tied to his teeth. He can't talk!"

Anton knocked on the counter top. The men stopped laughing and looked angrily at him. "*Ngo seung* (I want)," said Anton awkwardly in Cantonese, and pointed to some sweet and sour pork.

"The hide of him! Hitting our bench!" The workers began hitting various things, and pointing rudely at different types of food.

"All these A-Chans ever want is food. Let him wait!" They turned their backs and continued laughing and pointing.

Something inside Anton snapped. Perhaps his recently recognized skills gave him confidence, the courage to act. He had put up with this sort of rudeness so often since arriving in Hong Kong, but this was just once too often. He looked around, and saw the restaurant manager sitting in a small glass-walled office beside the entrance. Purposefully Anton strode over to the door and knocked loudly. The workers stopped laughing. This was not what they had expected. The manager looked up, saw Anton, and came out of his office.

"Sir," said Anton in English, "Why do your workers despise me? How do they know I come from Mainland China? I am as good a customer as anyone else."

He pointed to three insolent workers. "The lack of manners in your staff makes me angry! If you treat customers this way, your restaurant will collapse!"

The manager became extremely apologetic and yelled at the workers. Sullenly, they got Anton the sweet and sour pork.

"What else would you like, sir?" the manager asked with exaggerated politeness. "I am very sorry."

"I need nothing else," said Anton. "Here is your money."

Three months later Anton returned to this restaurant to get lunch. The service might be poor, but the food was excellent. It was the first time he had been back since the altercation. But to his surprise the door was locked, and the place empty!

"What happened to this restaurant?" Anton asked the proprietor of the neighboring shoe shop.

"Don't know." The man shrugged. "They just went out of business. It's a good location. I do well here. They made good food. But they just couldn't attract customers."

*****

Anton had felt great pride when he finally presented YuXiang with a sewing machine. In truth it was she who searched for the appropriate model and the shop that gave the best price. All he had to do was collect it and bring it home. He knew it was long overdue. But giving her the machine did not improve her health or alleviate his concern for her. Every day she coughed large quantities of bloodstained sputum. The new machine sat proudly on a table, but she rarely used it, complaining she was just too tired. The lung disease for which they had futilely sought help from countless doctors in Changchun and Beijing, was steadily getting worse. Anton studied his Chinese medical texts, tried acupuncture treatments, and boiled many potions for her. But nothing worked. He dared not voice to his concerns, but on the basis of his understanding of medical reports, he decided she could not live more than five years.

They visited many Hong Kong clinics, waiting hours in tedious queues, to no avail. Several times YuXiang was admitted for investigation, but she hated these episodes. Anton rarely visited her in hospital because, he said, he was too busy. She suspected, however, that he did not like the pall of sickness and death that hung over everything. Despite much talking and examining and testing the consensus of medical opinion was always the same: she had bleeding capillaries in her throat and trachea, but surgery to repair them could cost her life, or cause severe invalidism. Anton thought that since she was already an invalid, things could not be worse, but no one was willing to offer any treatment. Even a famous surgeon from Canada simply repeated what they had heard so often. It was a hopeless situation.

When Anton arrived home each day YuXiang was usually lying listlessly on the sofa reading cheap storybooks. He felt angry to see her so idle, but there was no sense in scolding. Doing the meticulous and beautiful technical drawing she had done in China required knowledge of either Cantonese or English, neither of which she had. Anton admitted she was better than he with her Cantonese, making good use of her trips to the market, but it was just market talk. She was not fluent enough to obtain work. Now she struggled to cope with housework.

Soon after he repaired Jan Van den Berg's violin, he returned from a teaching expedition to find her, as usual, reading. She sat up when he entered the room.

"You need to give me money," she said, looking directly at him. "I want two thousand Hong Kong dollars."

Anton was shocked speechless. Two thousand dollars! Two months' worth of family living expenses!

"I've been reading the story of Sanmao," YuXiang said, as though that explained everything.

Anton stared at her. She had spoken several times of this story about a Taiwanese woman composer. He was glad she was interested in a musical story. But it was, after all, just a story, not facts and science. And how did this story connect with needing so much money? I should not let her read these silly stories, he thought.

"I've been reading the story of Sanmao," YuXiang repeated. "I've learned something."

Anton raised his eyebrows, and shrugged. "Really?" he said.

"Sanmao says, 'Small sick find small doctor. Big sick find big doctor.'" YuXiang paused. "I want to do that. I want a big doctor."

Anton exploded. "But you have seen so many doctors! Big ones, little ones, green ones, purple ones and every other sort of doctor!"

"I've been asking people who is the best doctor," said YuXiang, ignoring his outburst. "The greatest doctor for my type of problem is Dr. Fosifo."

"Who? Where is he? Out on Lamma Island[5] or something!"

---

5. One of the larger of Hong Kong's many outlying islands, but with a low population.

"I have found his clinic. It's not far from here, and will be easy for me to get to. I can walk there. He does surgery at the Ruttonjee hospital. I want to see him. I have discovered it would take two thousand dollars."

Anton looked at his wife. She was beautiful, and totally supportive of him. During their tough financial struggles in Hong Kong not once had she complained of her own difficulties. She made good meals from cheap ingredients. She searched for appropriate schools for the girls. She did her best, he knew it, and he really cared about her.

"Yes, you can have two thousand dollars," he said, suddenly sitting down beside her and remembering the two months' salary she spent on him for *The Best Collection of Chinese Paintings*. "I am one hundred per cent behind you. I want you to get well."

YuXiang opened her mouth as though she expected to plead more to convince him, and closed it suddenly. There was absolute silence. Then she whispered, "Thank you." A tear slid down her face.

Next day Anton made a morning appointment with Dr. Fosifo so he could accompany YuXiang. His English was needed to give her history. When the day arrived, Dr. Fosifo listened carefully, and examined YuXiang meticulously. He left the room but was heard on a phone outside.

"This has been going on for ten years you say?" he said, returning and sitting behind his large desk. "You were treated for tuberculosis in Changchun?"

They nodded. "More than ten years," added Anton.

"Well, I think an operation can help this."

There was a shocked pause. Anton and YuXiang straightened their shoulders and stared at Dr. Fosifo, hardly daring to breathe. Anton had expected nothing. But YuXiang smiled, the faintest smile, but a triumphant one.

"Now, of course there are significant risks with surgery," Dr. Fosifo continued, "but I am willing to help. Are you willing for me to try? If you are, then I must ask you to sign this document."

Were they willing! After ten years they were willing to try anything. Anton leaned forward and signed the consent form a nurse brought in. In Hong Kong it is normal for a family member to sign consent forms. After all, how can a sick person be expected to know what is best, or be in a fit state to make decisions about their own health?

"I would like you to come into hospital tomorrow, Thursday," said Dr. Fosifo." We must do many tests to prepare you for surgery. If all goes well I will operate on Friday. Can you come tomorrow? I know it does not give much time."

"Tomorrow?" said YuXiang incredulously.

"I never thought my wife could be made well," Anton said, shaking his head. "We have seen so many doctors. I can hardly believe what you tell us!"

"It's a new procedure," said Dr. Fosifo. "You are not the only one with this problem, and we have discovered it can be treated."

"You have seen many doctors, I know," said the nurse, as she picked up the consent form. "Do you know how to get to Ruttonjee Hospital?"

"Yes, we do," Anton and YuXiang answered together. "What time should we be there?" added Anton.

"Eight o'clock."

"In the morning?"

"Yes."

Dr. Fosifo stood up.

"Tank u velly mush," said YuXiang slowly, as she left, bowing slightly towards the surgeon. He smiled broadly, but Anton was shocked. He had never heard his wife use English before.

They left the clinic in a daze. After all these years, and all the doctors seen, it was hard to believe something could be done. Because everything happened so quickly, they had no time to be anxious. YuXiang spent the day ensuring there was sufficient rice in the house, and telling NeeZi what to cook while she was away. Dr. Fosifo said she would be in hospital for two weeks, so the girls must take over household duties. They of course, were happy to do so, and told their mother to stop fussing. NeeZi was just a few days' from eighteen, and YanYan a mature twelve-year-old.

At seven o'clock next morning Anton hailed a taxi. They were on their way.

The Ruttonjee Hospital is in Wan Chai, east of Central Hong Kong and not far from where they lived. Originally established early in Hong Kong's history as the Seaman's Hospital, it later became the Royal Naval hospital, and was severely damaged during the Japanese occupation of Hong Kong. Soon after Hong Kong was returned to British rule Mr Jehangir Hormusjee Ruttonjee redeveloped the hospital as a tuberculosis sanatorium in memory of his daughter who died of tuberculosis during the Japanese occupation. It became the main center for the treatment of tuberculosis in Hong Kong. At that time there were large numbers of tuberculosis patients in the colony, and such a facility was badly needed. It became a world center for the management of this disease.

Within a few days YuXiang and Anton knew the operation had been successful. Although she could only manage a diet of congee she no longer coughed blood. A young Mandarin-speaking doctor took special interest in YuXiang. She learned it was not just the bloodstained sputum that was troublesome, but being kept awake at night with the irritating, troublesome cough. YuXiang admitted she was utterly exhausted and struggled to find joy in life.

"Hong Kong is a very polluted city," the young doctor observed. "Are you able to keep your apartment clean and dust free?"

YuXiang burst into feeble giggles. "Clean!" she squeaked, her voice still weak. "There is dust everywhere, wood shavings from my husband's work!"

The doctor raised her eyebrows. "He does that in your house?"

"He makes and repairs violins. There is always dust and wood shavings, no matter how hard I clean!"

"That sounds an interesting job. But I advise you find a proper workplace for him."

YuXiang shrugged helplessly. She had lived with wood dust for so long she hardly heard the doctor say that in addition to surgery they would give her medicine to help her feel better in herself.

But the treatments all worked. The sparkle returned to YuXiang's eyes, and she began reading magazines about life in Hong Kong, even though the traditional calligraphy was, at first, difficult for her to understand.

Anton's gratitude knew no bounds. For the first time since arriving in the British colony he was grateful they had come. After years of futile searching, at last his wife was on the way to health and happiness. He was immensely grateful to Dr. Fosifo and the Ruttonjee Hospital, and had difficulty wiping the beaming grin off his face when he thanked the staff for what they had done. He even told his mainland China friends there was something good about Hong Kong after all.

At the height of their joy, a letter arrived, one which should have caused rejoicing, but for its timing. As soon as he walked into her hospital room, on day six after her surgery, YuXiang knew there was a problem.

"What's the trouble?" she whispered, still hoarse from the surgery.

"Nothing, but, well, I hadn't expected this."

"What is it?" she whispered, fear in her eyes.

"It's NeeZi," began Anton.

"Oh no!"

Anton smiled. "Stop worrying! It's good, not bad, that has happened to her. She's been accepted to audition for the music conservatory in Vienna. She's been given an appointment just one week away. The letter must have been delayed, and taken longer than it should to reach us."

"You must go with her. You must go! I'll be fine. I don't need you while I'm in hospital! When I get home YanYan can help."

Anton looked at his wife. He loved his daughter and was very proud of her. But he both could not, and would not, leave his wife now. She had been through so much; this was no time for her to manage on her own. Perhaps it was a blessing from heaven that he literally could not go with his daughter. When he had applied for a passport for NeeZi a few months earlier he had also enquired again about a passport for himself. Because NeeZi had a Hong Kong student identity card there was no difficulty in her obtaining a British Overseas National (Hong Kong) passport. But he was firmly told he must wait for the completion of his seven years' residence in Hong Kong before he would be eligible for a passport. His Indonesian passport had long since expired, and he was always an "overseas Chinese" in China, not a citizen. He had not mentioned this problem to YuXiang earlier because he had not seriously believed NeeZi could study in Austria. But he had confidence in her, and was sure she could cope with going to Vienna on her own.

"I will arrange everything. I will talk to Tseng HsiangHo, and get his advice. Remember, his daughter has been to Vienna. NeeZi is a sensible girl. She will be fine."

"No!" persisted YuXiang in her most commanding whisper. "She is much too young to go so far alone! It would be terrible! What would people think of us allowing such a thing?"

Anton gazed at his wife. It would not be good to stress her when she had not completely recovered. "If I can't organize things properly I won't let her go," he said, temporizing. "I promise I will make sure everything is fine before I send her."

Anton made all arrangements, bought all tickets. The Tseng family proved most helpful, offering advice on hotels, public transport around Vienna, and reliable people to contact. NeeZi was given truckloads of advice from everyone, which she accepted patiently. The most notable advice she remembered from her father was she must never, ever, ever accept gifts from men, any men, any time. She had misgivings when she said goodbye to her tearful mother, still whispering, and still in her shapeless hospital gown. But she was determined to prove she could cope. She knew her father had taken a much longer trip (in time) when he was about the same age as she. She might not have many clothes in her small suitcase, but she had all the important documents in her handbag, and plenty of advice in her head.

YuXiang was home from hospital, a little frail after two weeks in bed with nothing much to do, but coughing no blood and feeling fine, when the phone rang. Anton's and YuXiang's heads banged together as they strained to hear.

"Mother! Father! Guess what! I've done it! I've done it! I've passed all the exams with honors! They've given me a full scholarship to study here! I can't believe it! It's a dream come true!" There was a pause in her rapid announcements. Her parents heard her take a deep breath, and suddenly there was a catch in her voice. "Thank you, thank you very, very much for allowing me to come. Thank you very much Father for all your teaching. Thank you for letting me come! And, Mother," there was another pause, and YuXiang heard a tiny sniff, "Mother, I am praying for you. I know God will get you well."

Anton's beaming smile said it all.

*****

Anton was feeling content as he walked along Des Voeux Rd in Central Hong Kong. After the toughest six years of his life, things were finally starting to improve, in fact, more than he had dared to dream.

First, YuXiang's surgery and her subsequent dramatic improvement was nothing less than a miracle. After more than ten years of suffering she was a new person. Her chronic, constant cough disturbed her sleep, made her listless, and without spirit. Now she was taking an interest in life, and again asking if there was some work she could do. She talked about learning to use a computer, and working from home. Yes, coming to Hong Kong, despite all the heartaches, had been worth it, for YuXiang's sake.

Second, NeeZi was on track to become a professional violinist. Whilst Anton had misgivings about this, he was joyful about her success. He knew how difficult the professional musical world was and thought it was not suitable for a young woman. But he had confidence she would cope in far away Austria. He smiled when he thought how the small violin made with love all those years ago in Changchun, had opened doors for both her and his success in music. Ah, so many things had happened since then!

The previous week he was invited to Lim KekTin's home to meet his famous brother, Lim KekTjiang. It was a wonderful evening. They reminisced about their lives in Indonesia, and their China experiences. After years of social isolation, it was wonderful to chat in such a relaxed way. Both Lim brothers assured Anton they would tell people he did violin repairs.

And now, here he was walking home carrying another violin to repair. He had made a careful examination of this violin and was sure he could fix it despite the owner's pessimism. But he needed to match wood to complete the restoration. Buying wood in Hong Kong was difficult, especially woods needed for instrument repair. Accidentally, however, Anton had found an unexpected supply of quality wood: rubbish dumps.

Many Hong Kong people were very superstitious. They strongly believed in *fung shui*. If, perhaps, their business was not as profitable as they would like, or there was a health issue in the family, a *fung shui* expert was consulted. Often this resulted in a diagnosis of poor furniture placement or offending architectural lines, and soon the upsetting and wrongly shaped furniture or walls were unceremoniously removed and dumped. Hong Kong had communal rubbish depots, where the garbage of whole apartment blocks was deposited ready for collection. These public garbage centers were treasure troves of goodies for anyone looking for something free. There was no law against sifting through this rubbish, and even European immigrants learned to make use of them at times. Anton's interest was always the discarded wooden furniture, and he regularly found pieces suitable for his repair needs at the garbage center.

This day he had looked carefully for discarded furniture as he walked home, but found nothing. Never mind, he could always look another day, and meanwhile open the violin and check its construction. Violinists often had trouble with their violins for years before they got them repaired, so usually they were not in a hurry. Well, except those who broke their instrument while on a concert tour.

Anton smiled as he remembered Benjamin Hudson. How could he ever forget this man?

The voice on the phone had been so distressed. "Sir, my beautiful, valuable, irreplaceable, Juanerius Gagliano violin is in pieces! Smashed! Ruined! What shall I do! I'm on concert tour: I'm the first violinist for the Columbia Quartet. But my violin is ruined!"

Anton at first had trouble understanding what had happened, but eventually pieced the story. Somehow, the clasp of the violin case had not been closed properly. As Mr. Hudson raced to the top of stairs behind stage at the Hong Kong City Hall, the case suddenly opened. The precious violin crashed down the whole flight of stairs, shattering at the bottom with an ugly long crack through its back.

Of course, Anton agreed to make the emergency repair.

It was a challenging task. Anton almost wished Jan Van den Berg had not recommended him so warmly to this very distressed musician. Van den Berg loaned his Tononi violin for the concert, but Hudson was desperate to have his Juanerius Gagliano with him when he returned to America the following week.

The midnight, the one a.m., and the all day oil was burned as Anton worked feverishly to repair this precious violin. But he succeeded. Benjamin Hudson flew out of Hong Kong with his violin completely restored. Later he sent a picture of himself and his quartet, duly signed with an appreciative message. It was a very satisfying episode to remember.

Anton turned into Li Yuen East Street, and headed down towards Queens Rd and home. Li Yuen Street was a popular and crowded part of Hong Kong. Narrow and steep, it was densely lined on both sides with a myriad of stalls selling women's clothing and accessories. Both locals and tourists loved it, and thronged its narrow confines. Anton was not interested in the stalls, merely using the street as a short cut, and was deep in thought about his repair of the Juanerius Gagliano violin.

Suddenly two men rudely blocked his way.

"Step aside!" they commanded.

Alarmed, Anton made to pass around these apparent crooks, but they grabbed him. Then one opened his hand and displayed his police ID. The plain clothes policeman forced him into a doorway off the crowded street. They took the broken violin and its case from him, and forced him to raise his hands above his head like a common criminal. Gentle, unassuming Anton died with shame as people stopped and stared.

One policeman systematically began frisking him as he searched the mortified Anton, while the other opened the violin case and began checking it carefully.

"Aieeyah!" exclaimed the policeman kneeling beside the case. "This guy is the wrong one!"

Angrily he gave Anton a shove, and hissed, "*Zhao*! (get lost)."

Anton attempted to talk to them. "Why did you stop me? What have I done?"

The policemen were impatient with his use of Mandarin, but they clearly understood. They shouted Cantonese at him, none of which he understood, except for one word, rubbish dump.

Burning with shame, Anton tried to melt into the crowd and get home as fast as possible. The police made no attempt to apologize for publically embarrassing him so badly.

YuXiang was kind and sympathetic when he poured out his tale of woe, but had no advice except that perhaps he should not go snooping around garbage dumps. However, Anton pointed out that this activity was not illegal, and it was probably just bad luck he had been stopped. When his daughters learned what had happened, although they were initially kind and sympathetic about their father's terrible disgrace, they joked that father had always been looking for things along the road, ever since he went hunting for chicken food as a boy in Indonesia.

Sadly, this was not the last encounter with police. Several months later Anton struggled to carry home three violins. He was not complaining. It was proof he was becoming recognized for his repair work.

But just as he started to cross the road to his home, two uniformed police forced him to put down the violins, once again raise his hands in the air, and again body frisked him and searched his violin cases. This time they were not satisfied, and demanded he take them to his home to search it.

Horrified, Anton led them across the street, and tried to sound cheerful as he called to YuXiang to warn her.

"I have guests who want to look at my work!" he shouted.

YuXiang came forward smiling, but seeing her husband's strained appearance, and the nature of his guests, recognized he was in trouble again. She melted into the wall.

The police, both extremely young, searched everything in the Sie's apartment, and eventually agreed this was indeed a genuine violin repair workshop.

"*Mo si* (no problem)," said one, turning towards the door. His mate quickly followed him, and they left, banging the door behind them.

Anton was furious, but what could he do?

"It's terrible the way they chase me!" he exploded as the door slammed shut behind the police. "What can I do to stop it?"

"Do you think it's because we speak Mandarin?" asked YuXiang.

"Yes, I'm sure they recognize I am mainland Chinese. But I was not talking to anyone on either of these occasions. How do they know where I come from?"

YuXiang was silent for a long while.

"You know, I don't think you will like what I say, but I think it's the way you dress. I know you've never been interested in clothes. I like you for that. I can laugh about all the times you went out with shoes that did not match! But perhaps you should try to dress a little more like people here in Hong Kong."

Anton looked utterly bewildered. "Whatever do you mean?" he asked.

"Well, have you bought new clothes since you've been here?"

"I bought two shirts when we first arrived!" he said defensively. "But there are so many other things to spend money on. Anyway, what's wrong with what I'm wearing?"

YuXiang smiled. "There's nothing wrong with your clothes, but they're just not what locals wear. They tell everyone that you come from China."

Anton looked down at his offending trousers. They looked perfectly fine to him. Whatever was his woman complaining about?

"You're a smart person," YuXiang persisted. "Haven't you noticed that if you use sharp eyes you can see the difference between a China Chinese person, and a Hong Kong Chinese? Hong Kong business people dress according to English style of clothing. Very smart, you know, a flower in their jacket pocket! Hong Kong workers dress according to American cowboy fashion. They're both very different from the way we dressed in China. For example, have you seen anyone here wearing a Mao suit or jacket?"

"A Mao jacket? Well, no, not exactly," conceded Anton.

"That's it! Tell me," continued YuXiang warming to her topic, "could you afford, now, to buy some new clothes?"

"Well, maybe, possibly, perhaps," he said thoughtfully. "But I want wood, and tools, and business cards..."

"Exactly! Why don't we go out this afternoon, right now, and get you a few new clothes? Not too many," she added, when she saw how horrified her husband looked. But sensing her opportunity, she declared, "Let's go, right now!"

Anton had never before noticed the plethora of clothing shops in Hong Kong. He rarely noticed any shops, to be sure. Even at the best of times, his mind was always on more serious issues. Clothes were simply not worth thinking about. Now, as YuXiang marched him to the closest menswear store he suddenly realized just how different his clothing was compared to the garments in these stores.

"I can't wear things like that!" he exclaimed as he entered the shop. There were smart three-piece suits, and faded navy cotton pants only suitable for a cowboy. "How could anyone wear such dreadful clothes!"

YuXiang giggled. "Not everything is like those jeans," she said, gesturing towards the offending trousers. "There will be clothes you feel comfortable wearing."

Half an hour later Anton left the shop, the proud owner of two pairs of dark grey gabardine trousers, three very pale blue shirts, and a simple charcoal-colored jacket. YuXiang had a very self-satisfied look on her face. At last she had been able to get Anton to buy new clothes!

But the police harassment did not stop. Anton discovered that if you walk slowly, looking around for pieces of discarded wood, police will definitely notice you. Despite the beautiful new clothes, a few weeks later Anton was once again stopped by police. Anger swelled inside him, but he kept calm. This time he used a trump card.

"What should I do for you, sir," he said in English, as the police officer roughly pulled him aside.

Suddenly the policeman changed his belligerent, demanding attitude.

"Oh! Oh! Sorry sir," the young officer said, turned on his heel, and left.

It was not the last time Anton was accosted by police. But by using English, in which, with his teaching, he was becoming ever more proficient, he was never again humiliated by a body check or a house search.

"It makes me so angry!" he grumbled to YuXiang after another encounter with law enforcement officers. "If I speak Mandarin I am a low class person, but if I speak English suddenly I become a good person, a high class citizen! It is not right that there is such discrimination."

"No, it's not right," she agreed. "But was it right in Changchun when you rode a bus to work and I had to walk? And this colony has made me well."

Anton opened his mouth, and closed it quickly. He stared at his wife in silence for several minutes.

"NeeZi is following her dream," he conceded. "And I am glad you are well."

After a few minutes he suddenly declared, "You know this place is interested in one thing only. Business. No business, no friend."

YuXiang shrugged. "So?"

The idea of business worked on Anton's brain. In China he did not have to tell people what he did. He was a respected member of the university faculty. But what was he, who was he, here in Hong Kong? What legitimate occupation could he claim?

A few months later he applied for, and was granted, business registration as a teacher of music and repairer of violins.

"Now that," he said to himself as he gazed at the official document notifying him that he was legally allowed to operate his teaching and repair business in Hong Kong, "that should make me belong." He smiled to himself. "But I think it's my students, and the violinists I help that count, not a business registration!"

*****

"There's a letter for you, from Indonesia," YuXiang said, as he arrived home after a pleasant afternoon teaching at Lam Tei.

"Oh, that's good," Anton replied absently, and went to his workbench. He had a difficult repair and had just conceived how to approach the problem.

"It has 'urgent' written on it," said YuXiang. "YanYan told me."

"Urgent," echoed Anton absently, "Urgent. I wonder why? They have never done that before."

"You should look at it," said YuXiang firmly.

Anton reluctantly put down his chisel, and picked up the letter.

He scanned its pages, and suddenly gave a heart-rending cry. "Oh no! Oh no! Oh no!"

"What is it?" asked YuXiang in alarm.

"My mother! Oh no! My mother! She died two weeks ago. Heart attack." Anton dropped the letter on the table.

"Can you go to the funeral?"

Anton spun angrily. "Funeral? What funeral! It's gone, gone! Done! Go? Anywhere? With no passport? Two more years, at least, if ever, before I get one! What are you talking about!"

"I'm so sorry," YuXiang answered softly. "So very sorry."

"Oh, mother," lamented Anton, turning roughly from his wife as he gazed sightlessly at the letter on the table. "I never thought when I left home thirty years ago I would never see you again. Oh mother! I meant to do what was right. I really meant to do what was right, I truly did! I tried to honor you."

He paused. There was a long, long silence. "At least Mother knew I am recognized as a violin maker and repairer in Hong Kong."

He laid down his chisel, and without another word walked out the door. Two hours later he returned, wordlessly picked up his chisel and began work on the violin.

The pain of loss, the frustration that he could not even attend his mother's funeral, weighed heavily on Anton's mind for weeks. He retreated into silence, and work, hard work, lots of work. Students were taught, violins were repaired, but his wife and daughter were ignored. YanYan dreaded the times she had to call her father for meals. His eyes would flash, and he would bark: "Can't you see I'm busy?"

"Husband," said YuXiang gently one morning after he had refused yet again to eat his evening meal, and sat staring at a local Chinese paper. "You did your very best for your mother. You could not do more. Surely that is what a truly filial son should do? Don't blame yourself for the bad things others do to you. You have a good family here, and I know your mother was very proud of you. Please remember you are a husband and father."

"Yes, yes," said Anton. "But coming here now seems pointless."

"Things are getting better, you must agree. My health has improved. NeeZi is studying music with that famous school in Austria. I don't think she would have done that if we had stayed in China. YanYan is doing well at school."

Anton nodded, slowly and thoughtfully. "Perhaps."

"And how many years have we been here now? Almost six, isn't it? It won't be long before you can get a passport and see your father."

Anton folded his paper, and sat staring at the wall. Finally, he stood up. "What did you say you made for breakfast?" he asked, and smiled, the first time for several weeks.

Sketch of Chinese dancers, an example of Anton's art.

Unfortunately, without color reproduction, some of the beauty of Anton's paintings is lost, but their general content can still be appreciated. *Stradivarius* is one of his most loved paintings.

Anton's richly colored painting, *Flamenco.*

Anton's whimsical painting, *Maestro in the sky.*

Anton's playful painting *African Grey leads the Quartet.*

Anton's painting *Paganini.*

Anton's regal portrait of Anne Sophie-Mutter.

Press photo of NeeZi during an encore presentation of her winning violin performance, at Yuen Long District Hall, 1977.

A very happy Anton presenting Jan Van den Berg with his repaired Tononi violin.

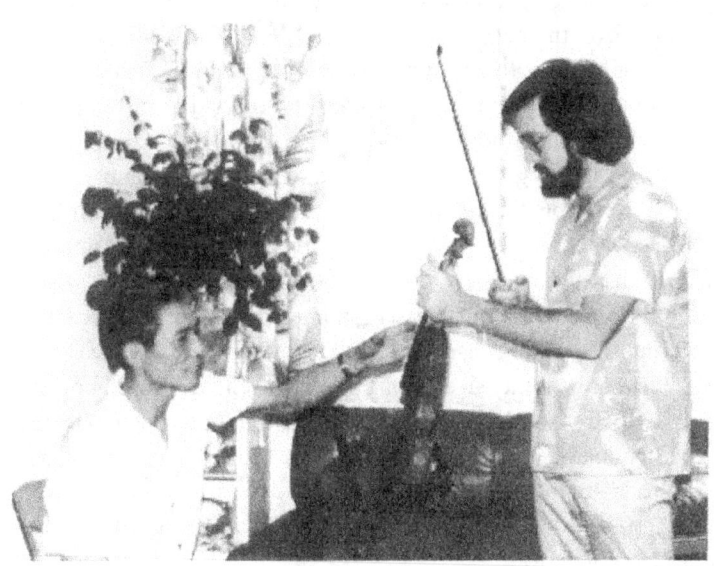
Anton examining the broken violin of Benjamin Hudson.

Fourth Movement

Violin Expert

"This is a beautiful violin," said Anton, as he picked up the shining chestnut instrument from the coffee table in the reception room of the Hong Kong Philharmonic Orchestra offices. Carefully he passed it to Carl Pini, concertmaster of the orchestra. Pini drew his bow across the strings and nodded contentedly.

"Mr. Sie, I'm so glad you've been able to fix it," he said.

"Would you like to *really* test your violin, sir, and make sure I've done a good job?" asked Anton, his eyes twinkling. "I could accompany you on that guitar over there."

Pini looked up, his eyebrows raised in surprise. "You play the guitar?"

Pini was not in a hurry, and Anton's unexpected offer sounded fun. He chose a simple piece, not wishing to cause Anton loss of face. But Anton, instantly recognizing the melody, played a masterful accompaniment. Pini played a more complicated piece, but Anton played along with it competently. For half an hour the two men were immersed in the simple fun of making music.

"Of course you know this is a genuine Carl Bergonzi instrument," Pini said proudly, lowering his violin. "I bought it in 1960 for relatively little when I was only twenty-five. I don't think the vendor realized how valuable it was. Actually I think this violin was the reason I was offered the lead role in the London String Quartet. Ah, that was fun! Did you know Bergonzi learnt his violin making from Stradivarius?"

"Yes, I did know that," replied Anton. "Your violin must be worth a great deal more now."

"Really? You know that Bergonzi learned from Stradivarius? Ah, but it bothers me that I don't know which Bergonzi I own!"

"Oh, I can tell you that," responded Anton. "When I was repairing it I checked that out. You see, I learn a lot about these old masterpieces when I repair them. I repair and learn at the same time."

"Did really you check which one it is? Can you tell me?" Carl Pini's eyebrows again shot up quizzically. "For more than twenty years I've tried to find out!"

"Let me show you." Anton retrieved his well-worn copy of Carl Jalovec's book of violins from his briefcase. "Of course, I myself was interested which Bergonzi it was." He opened his book to the place marked, and laid it across the coffee table. "See, it's the 1732 Bergonzi. See these two parallel thin patches on the upper left bout? They prove it." Anton pointed to the picture, the small identifying patches, and the corresponding marks on Pini's violin.

"These?" Carl Pini exclaimed. He peered at his violin as though he had never seen it before. "You mean these little tiny things that look like scratches?"

Anton nodded. "The wood must have had a small defect due to a worm, and Bergonzi cut the bad wood out and repaired it."

"Well!" responded Pini, staring hard at his violin again. "Well, fancy that! After all these years I finally know what I've got! That means I'm doubly grateful to you! You not only repaired my violin, but you identified it as well."

"Happy to help," said Anton.

There was a pause, and Pini asked, "How did you start repairing violins? What got you interested?"

Anton explained his childhood interest in music, his training in physics, the years of spare time study in China, and the tiny violin that won the attention of his daughter's examiners.

"You mean you make violins? By hand? You still do that?"

Anton nodded.

"Really? How long does it take you?"

Anton was silent for a few moments. "It's a delicate process, with many steps. I have to wait for things like glues and varnishes to dry. But if I work four hours a day I can make a violin in three months. My process is a little faster than the Cremona luthiers, but I still can't make more than four violins a year."

"Just four a year!" Carl Pini whistled, and shook his head in wonder. He lifted his violin from his lap and looked at it with admiration. "I've always thought this instrument was special. Now I realize just how special it really is!"

Anton smiled and bowed his head deferentially.

"Excuse me for being personal, sir," Anton observed, pausing diffidently, "but I can't help noticing that you have a swollen right thumb."

"Oh, I've had that for years, probably at least ten. No doubt poor technique when I was young. I've used heaps of anti-inflammatories, but they made no difference."

"Maybe I can help you, sir. A Chinese medicine called *Pianzihuang* could help. You can take it orally, or rub it on the swollen thumb."

"And it works? Really? Honestly? Ah, well, why not? I'll give it a try. Thanks very much." Pini grinned.

"I'll come back in a couple of days and give you some."

A few days later Pini accepted the medication graciously, and later told Anton it was beneficial.

Anton heard no more from Carl Pini for several months. Then one day the phone rang, with a very distressed Pini on the line.

"Mr. Sie! Mr. Sie! I need your help! I urgently need a guitarist! I can't get anyone to work with me!" Carl Pini was breathless with frustration and concern.

Anton suggested several professional guitarists he knew of, and Pini agreed to contact those he had not already called. But a few hours later he was back on the phone just as distressed.

"I've called all the guitarists you told me about, but none of them agrees to work with me. The problem is the same with everyone I called. They all claim there's not enough time to learn the music of the Don Pasquale opera. Please, will you to work with me? I know you can play classic guitar. I still remember that happy afternoon we had when you identified my violin for me! If you won't help me, then the performance of Donizetti's marvelous opera will be cancelled. That would be terrible! A guitarist is essential for the production."

There was a long silence. Anton suddenly became aware of his shama singing with abandoned gaiety. He made up his mind.

"Sir, I'm not a professional guitarist. I've never done anything like this." He paused, and could hear Pini's heavy breathing. "But . . . but . . . if you really want me . . . I will try to help you. I'll do my best for you!"

There was more heavy breathing on the phone, as though Pini was trying to believe what he had heard.

"You mean you will help me?" A heavy sigh.

"Yes."

"Oh, I can't thank you enough! You've saved the day! I'll get the music to you immediately."

"Yes, that would be good. As you say, there's not a lot of time for me to perfect it."

Pini was as good as his word, and the opera score arrived in Anton's mailbox that afternoon. Two days later Anton received an official letter inviting him to be guest guitarist with the Hong Kong Philharmonic Orchestra for the three performances of the opera.

Anton stopped everything except his teaching, even his violin repairs. He spent every moment practicing the guitar pieces for this famous comic opera. Very quickly he discovered why other guitarists were not willing to take the role: in the second act the tenor has an aria accompanied only by a single guitar. But Anton enjoyed playing this serenade with tenor Brown Bradley, and the rehearsals went well. With Anton agreeing to take the lead guitar role, Pini had managed to find another young guitarist, Lo ChiKeung, who was willing to assist, although ChiKeung insisted Anton play the solo accompaniment. The three days of performance were most enjoyable, and the tenor soloist appreciative of Anton's accompaniment.

The only reason Anton agreed to perform was to help his friend. He was proud to hear Pini say, "Mr. Sie, you're a real artist! Not only have you repaired my violin, but you're a performer as well! You've been such a help to me. I am very grateful."

But when a check for nine thousand Hong Kong dollars arrived from the Hong Kong Symphony Orchestra management, the whole episode became a dream. Why, it seemed only yesterday that Anton was paid a paltry thirty dollars for paintings, not earning enough for his family to even eat. Unbelievably, in just three days of performance he earned nine months' worth of the family's frugal living expenses! He found himself intensely hoping his Ocean Painting Studio art shop boss attended operas.

It was time for celebration, again. But this time Anton decided to be self-indulgent. His love for birds was as strong as ever, and the shama gave his family so much pleasure. Why not buy another bird?

It was only a small detour on his way home from teaching to go to bird market in Hong Lok Street in Mong Kok, Kowloon. The tinkling cacophony of birdcalls was as sweet as ever to Anton's ears. But nothing could sing better than a shama, and he could not decide what bird to purchase. Then he passed an unprepossessing-looking large grey parrot with a tail splashed red. Aha! An African grey! He had heard of these birds, said to be as intelligent as a child of five years. Their ability to mimic human speech was renowned, they could learn numbers, and challenged the idea that only primates have intellectual function.

Anton walked up and down the street, oblivious to all other birds. Yes, the African grey was expensive, very, but it would be so much fun to have this bird around his home. He kept walking, trying to be the sensible and logical scientist, but finally admitted he liked the idea of sharing life with an intelligent bird. He turned back and bought the African grey parrot.

*****

On a clear spring day, very welcome after many days of clinging winter fog, the setting sun sent long sharp pedestrian shadows across the footpaths and up the walls of Hong Kong Central skyscrapers. Anton walked home along bustling Queen Street, his heart content. His plan of teaching students in their homes at the residential apartment block of Tsuen Wan Adventist Hospital was going well. The students were responsive, courteous and progressing steadily. The parents treated him as a respected person. As he turned into his apartment building he stopped to collect mail from the bank of boxes by the front door. He smiled when he noted a Chinese stamp on one letter.

"Must read that first!" he thought.

YuXiang was busy cooking the evening meal. Since her surgery at the Ruttonjee Hospital the terrible lethargy of many years had left her, and she was a happy, productive woman. She was always busy, although Anton was not quite sure what was occupying her. After briefly greeting her, he went to his workbench and opened his letter.

To his surprise, it was from Wang Xiang of the Musical Research faculty of Beijing. Anton had not heard from him since he had moved to Hong Kong six years earlier. Suddenly he let out a long low whistle.

YuXiang turned sharply, anxiety shadowing her face. But one glimpse of her husband's beaming smile showed it was not bad news.

"Well!" said Anton, pleasure beaming from his eyes. "Well! Would you believe it! They're begging me, did you hear, begging me, to go to Beijing to help them with translation duties for the visit of Carleen Hutchins and her husband Norton! Carleen Hutchins, no other, can you believe! They think I am the only person who knows enough of both English and the acoustics of stringed instruments to help with translation. The Hutchins will be in China in a few months, and give lectures in Beijing, Shanghai, and Guangzhou."

Anton paced backwards and forwards across the floor, rereading the letter to make sure he was not dreaming. "Not in my wildest fantasies!" he muttered. "This is wonderful!"

"Who are those people? Do you know them?" asked YuXiang, puzzled at his excitement.

Anton grinned. "Know them, no. Know about them? Do I ever! She's the most noted violin maker and researcher in the world today. I learned about her when we were in Changchun, and read everything I could find about her in the library. When we got here I even paid for a mail order copy of *Scientific American* because it contained a paper she had written."

Seeing YuXiang's eyebrows shoot up he added apologetically, "That was before I knew how hard it was going to be to get work. It didn't cost that much."

"Really?" said YuXiang, smiling mischievously.

He took a few more paces. "I believed her first paper on violin acoustics was published in that magazine. I found articles on her work in copies of the *Journal of the Cat Gut Society of America* in the library here, and I've corresponded with her since we got here. She does superb work, but she claims her studies show something quite different from what I found. I wrote to her about it. A few months ago I received a polite reply, but it did not answer my question. Wow! Now I can talk to her personally!"

"Oh, Anton, this is a marvelous opportunity for you!" YuXiang smiled and turned back to her cooking.

Anton was lost in thought. Yes, he had written to Carleen Hutchins after he arrived in Hong Kong, sharing his research done in Changchun on violin acoustics. He was concerned that his study gave the opposite to what Hutchins claimed she had found for the best plate tuning. He believed she would have the best research facilities available in the USA and therefore should be right. He admitted his own facilities in China were very limited. But what he lacked in equipment, he made up in meticulous care. Moreover, he was able to make excellent violins using his own conclusions. Therefore, the difference between their results regarding the *Eigen* frequencies of plate tuning was a matter of serious concern to him. He never expected to have an opportunity to speak to Mrs. Hutchins in person.

As he cheerfully wrote an acceptance to his invitation he was sure this was the chance, *the* big chance, of his life. His gratitude to the authorities in China for giving him this break was unbounded, and he was careful his letter expressed this. He reviewed all he had ever heard about Carleen Hutchins and her work, and all his own research.

*****

Anton remembered Mrs. Hutchins had begun her musical life as a trumpeter, and only switched to a viola after she was a schoolteacher in order to join a small music group organized by a fellow schoolteacher Helen Rice. The cheap viola she bought did not please her, and much to her colleagues' disgust she decided to make a better one herself.

Her friend Helen Rice introduced her to Frederick Saunders, a physicist who had been researching violin acoustics for twenty years, and she joined him in his research. Saunders was reluctant to study the violin plates (the back and front of a violin), but Carleen convinced him to study plates as well as other parts of the violin. She used children's decorative glitter to visualize the acoustic movement of the plates. Hutchins' and Saunders' research seemed to confirm the conclusion that the best violins had a top plate with a tap tone (*Eigen* frequency) one semitone higher than the back. Anton challenged this conclusion, as his research indicated the top and back plates should have at least one *matching* tap tone, and preferably two or three.

Anton was interested in one of Carleen Hutchins' most fascinating innovations: the development of the violin octet. Henry Brant, an American composer noted for unusual instrumental combinations, asked Hutchins to build an ensemble of violins to create a continuous range of violin tone. They originally planned to add seven instruments to a conventional violin, but this proved difficult. A *mezzoviolin* (which is two and one half centimeters longer but with narrower ribs) covered the same tonal range as a conventional violin, but blended better with the other instruments, and so replaced the regular violin. Interestingly, additional members of the string family were in common use until the end of baroque music period (approximately 1600–1750). J. S. Bach composed for a *violoncello piccolo* and a *violin piccolo*.

One assertion Hutchins made which Anton had taken special note of, was research indicates it is possible to make a fine violin every time, with no need to pay millions of dollars for a good instrument, although it can take from twenty to eighty years for a good player using an instrument consistently to properly season a violin.

*****

In the summer of 1982, guest of the Chinese government, Anton travelled to Beijing by air. It was his first airplane flight, and he was enchanted. YuXiang persuaded him that new clothes were essential for this trip, and for once he made no protest. He even submitted to buying some blue jeans that YanYan confidently assured him would not only make him acceptable to the American guests, but help them feel at ease. He decided a white shirt with a striking blue dragon pattern would match the jeans, which he was surprised to discover were quite comfortable to wear. He could not be parted from the black jacket he had worn so often to present his repair work to members of the Hong Kong orchestra, and secretly YuXiang agreed it looked well on him.

But as Anton was being driven from Beijing airport his excitement at meeting the famous violin expert was far greater than his interest in his first plane flight. Apart from the personal opportunity it gave him, he also realized his responsibility. If he could not give a clear translation of her lectures, the academics in China would not learn the significance of both her work, and eventually Anton's own.

From the moment they met the two dedicated violin researchers established a warm and friendly relationship. Anton quickly understood why so many people had been willing to work with this generous-minded and unassuming woman. Carleen Hutchins was warm-hearted, not only interested in all aspects of violinmaking, but also the achievements of others. Anton did not at first broach the subject of the difference between his research on plate tuning and hers, but eventually he was brave enough to do so. Although skeptical, she indicated she was willing to look into the matter. They agreed they would correspond on the subject.

The lecture tour began the day after Anton's arrival. In Beijing it was held in the Academica Sinica Acoustic Hall, a huge auditorium that could seat up to three thousand people. Department leaders and senior researchers in music and acoustic departments from all around China attended. Reminding him of his arrival in China almost three decades before, long speeches began the program. Professor Ma DaYou made a brief introductory speech, then Professor Chen Qiang spoke at length, followed by welcoming words from none other than Anton's old friend Dr. Wang Xiang. Finally, Carleen Hutchins rose and lectured on the history of the general research being made into the acoustics of musical instruments, and Anton did his best to translate accurately. Mrs. Hutchins was not used to working with a translator, and spoke rapidly for long stretches without a pause. It was challenging work for Anton. After a short break she gave another lecture explaining her methods for consistently making high quality violins, and the ideas behind the violin octet. The second lecture proceeded more smoothly because Anton asked her to pause more frequently. After his wife's speeches, Morton Hutchins gave a lecture on the issues and problems surrounding wood measurement. The lectures were given in a lively style, and the applause was long and sincere. Chinese interest in this first opportunity to hear from the world's best-known violin researcher was obvious. Only years later did Anton discover that the written translation of the lectures was not as accurate as he would like.

Anton was grateful for this opportunity to revise his own understanding of violin construction. Violin making reached a peak in the Antonio Stradivarius (1644–1737) era, and since then the art and method of crafting quality violins had apparently declined. In the nineteenth century studies into violin acoustics began with the French physicist Felix Savart and German physicist Ernst Chladni (1756–1827), and were continued by the violin maker Jean-Baptiste Vuillaume (1798–1875), pianist Backhaus, engineer and acoustician Lothar Cremer (1905–1990), and to some extent Julius Lothar Meyer (1830–1895). They studied whole body violin vibration, but since none of them obtained consistent results the work stopped. Because of this failure it was thought impossible to discover the secrets of the Cremona luthiers. Saunders, Hutchins, and Sie however, studied not only whole violins, but also, and particularly, the vibrations of violin plates. Hutchins published her findings in 1966, and unknowing and completely independently, Anton Sie finished his studies in Changchun in 1968. Much of their work had correspondence except for the critical issue of tuning the plates.

After the successful lectures in Beijing, Anton and the Hutchins spent a day sightseeing. They visited the Great Wall and Forbidden City. The party then went to Shanghai where the lectures were repeated, also with great interest and response. After the Shanghai lectures they once more had opportunity for relaxation, and this time went south to Hangzhou and famous West Lake. Anton remembered Carleen Hutchins bought a blue and white patterned blouse she liked and subsequently wore frequently for many years.

The final stop on the China tour was Guangzhou, at that time the Chinese center for mass-produced student violins. This lecture stimulated a lively question and answer session on plate tuning. Anton found it hard to keep silent on this issue, but out of courtesy to the American guests, he did not mention his own research at this time. He was extremely disappointed that Carleen Hutchins appeared to show no interest in his work, nor to appreciate his findings. Finally, Canton television made a report on the Hutchins' visit, and the city honored the visitors with a violin concert.

After the China trip Anton and the Hutchins entered Hong Kong and repeated the lectures at ChungChi College, which later became the Hong Kong Chinese University in Shatin. When his work of translating was finished, Anton became tour guide showing the Hutchins the sights of Hong Kong.

Anton was delighted when Carleen and Morton Hutchins asked to visit his home and was surprised and thrilled when she indicated she was keen to try his handmade violins. He dared hope their quality would convince her of the truth of his research. When she played his violins she agreed they were of extremely good quality. To his delight, although she remained cautious about accepting the results of his plate tuning studies, she was sufficiently impressed with his violins to encourage him to write a paper regarding his research and its conclusions for the *Journal of the Cat Gut Acoustical Society*, which he later did. To his surprise, and immense joy, she also generously invited him to become a member of the Catgut Acoustical

Society. Before she left Hong Kong she took a charming photo of Anton and YuXiang.

As Anton bid the Hutchins farewell, he dared dream that he could continue to work productively with this dedicated Western luthier.

*****

A few years after Anton began his repair work for members of the orchestra in Hong Kong, Sandra Wagstaff, an instrument merchant from Brussels, opened a musical instrument shop on Wyndham Road near where Anton lived. She advertised extensively, and made contact with Hong Kong music associations including the Philharmonic School. She used every method she could to get herself established.

Anton quietly observed her actions, and made no contact with her. After all, he was more interested in perfecting the quality of his own violins than selling European instruments. One day Sandra Wagstaff advertised that she would hold an auction of expensive Italian violins. Unfortunately, she could not authenticate her instruments. But, a resourceful woman, she discovered there was someone in Hong Kong who had what she needed to remove this deficiency.

When Anton picked up his phone one morning he was surprised to hear a European woman with a strong accent speaking. What she asked for astonished him.

"No madam, I am not offering any of my violin books for sale."

There was a great deal of rapid talking.

"Yes, I understand you need them to authenticate violins that you have for sale, but I am not a bookshop."

More rapid talking.

"Yes, you may visit me, and we can discuss this further if you wish. But let me insist, I am not interested in selling my books, no matter how much you may offer to pay for them."

A few days later Sandra Wagstaff arrived at his home. She was polite and did not appear as demanding as she had sounded on the phone.

"This book is amazing!" she exclaimed as she opened his prized volume of Carl Jalovec. "This would be all I need to let people know my violins are authentic! Can I take it?"

Anton took a deep breath and once more insisted he was not a bookseller.

"Then can I borrow it for a few days?"

"Madam, I am not a library! You simply do not understand! What I am is a master luthier!"

Mrs. Wagstaff's manner suddenly changed. "Of course," she said deferentially. "May I ask you, as a special kindness to me, if you would allow me to borrow your beautiful book for just one week. Only one week! I will take special care of it, and make sure it is insured for the week it is with me."

Anton took a deep breath. Jalovec was not only a prized possession; it was also a special gift. But he could never resist being kind, or helping someone in need.

"That book is very valuable to me, madam. I need it for my work. It was a special gift from a very dear friend. But I am willing to allow you to use it for one week. Please honor my trust."

"Oh, thank you! Thank you! How can I thank you enough!"

Anton had very serious misgivings as his two precious violin books were carried out the door in Sandra Wagstaff's voluminous shopping bag, but in his heart he was confident he had done the right thing.

Many of his friends thought he was incredibly foolish to have agreed to Wagstaff's requests. They were sure Sandra Wagstaff's sale, by flooding the Hong Kong market with Italian violins, would undermine his livelihood as a luthier. But Anton was always willing to help, despite his concern regarding the possible loss of his precious books. He was greatly relieved when Wagstaff returned the books the day after her auction. His trust had been vindicated and he was grateful she was responsible.

Later, however, Anton realized that helping Sandra Wagstaff meant he had actually helped himself. As people attending her auction referred to the large authoritative books lying on her table (to authenticate the instruments they were buying) they could clearly see she had utilized the expertise of an Anton Sie. Instead of harming him, assisting her became a huge advertisement for himself. People learned there was violin expert in Hong Kong, someone Sandra Wagstaff had relied on, and his name was Anton Sie!

When these musically interested people recognized they had a world-class violin maker and repairer in their midst, they became his customers, and Anton thus met, among many others, Dr. Watson, Principal of the Academy of the Performing Arts, civil servant John Boyd, and the wife of a governor of Hong Kong who was a fine amateur violinist. Eventually these contacts lead to his being invited to join a chamber music orchestra that played the Bach Brandenburg concertos at the New Year Concerts in Government House, around the year 1990. And all because he helped someone others thought would harm him!

As Anton flipped through his safely returned books he recognized science and music are like a relay race. Winning the race does not consist of one single person's performance, and passing on the baton is essential to achieve success. His own masterpieces were built on the work of luthiers who had perfected the art of making great violins. He was more than willing to help others as they too passed on the baton of musicality.

Anton strongly refutes suggestions that he might develop hero status. Heroes, he says, can be very selfish, demanding honor and recognition for themselves. When people express appreciation for help Anton has given them, he reminds them that their success is at least seventy per cent due to their own hard work. He might have given some encouragement, some hint, but unless they had followed it, the hint would have meant nothing. He suggests people should recognize what he calls "sincerity and eternity." Without a true and sincere attitude to life and others there can be no real success, and without appreciation for eternity even success is only temporary. Carleen Hutchins succeeded because she collaborated and co-operated with many other people, and so, he believes, has been his own case.

Anton cites an example of these principles. One day at a violin shop in the city he met a man called Cho WaiLap. Cho had an old violin and played very well. However, at the time of their meeting Cho was working as the night cleaner in a large department store. It was demanding, demeaning, and worse, very low paid work. Cho was miserable and discouraged. He offered his violin to Anton, who, however, recognized that although Cho owned only a cheap violin, he was a good musician.

"But sir, you play many good violin pieces well. Why give away your violin?" asked Anton.

"Look, I've tried and tried to get better work to use my music, but this cleaning is all I can find. No one wants my violin playing, and I need money urgently."

"But you love making music! Why not work to develop it? Music is so delightful! With music you can share beauty, and friendship, and bring peace to others! Don't give up your music! Keep on developing your skill. Try getting a British diploma." Anton smiled to hear himself giving advice that he had so desperately needed when he first came to Hong Kong.

Dubiously, Cho agreed to try. It was not easy. Four times, again and again, he failed the English diploma of teaching examination. But he took Anton's advice and did not give up. Whenever Anton met Cho he encouraged him to keep on trying. On the fifth try he was successful. With this certification he was able to get a good position as a violin tutor with the well-known Tom Lee Music Centre of Hong Kong, and continued there for many years.

Anton acknowledges that he encouraged Cho to keep on trying, but he takes no credit for his success, which was due entirely to Cho's own hard work, and the blessing of "eternity".

Anton's own optimistic commitment to hard work is illustrated by the incredible story of Chou HuiRen. One day a distraught and very sad Mr. Chou arrived at Anton's house with a bag containing a shapeless, lumpy, newspaper-wrapped parcel. Mystified, Anton unwrapped the bundle, and was astonished to discover about two dozen pieces of broken wood, some of which had obviously once been part of a violin.

"How did this happen?" exclaimed Anton, shaking his head in bewilderment, and stirring the jumbled pieces of wood with his finger.

"You won't believe it," Mr. Chou replied, "but my little boy did this."

"Your boy? A mere child broke this violin?"

"Yes, he was watching a *kung-fu* program on television. He got very excited by the exploits of the hero, and decided he needed a weapon to demolish his enemies, like the hero did. He looked around for something big he could hit with, and as I said, believe it or not, saw my violin in the corner of the room, and used it. My wife heard him shouting and banging, but did not realize what he was doing. I must say he was very upset when he accidently smashed the corner of our cabinet, and the violin broke into so many pieces. Then he started screaming, and that's when my wife discovered what he had done!"

Anton opened the parcel wide, trying to collect his thoughts while bizarre visions of a small boy brandishing a wicked and dangerous violin ran through his head. Idly he counted the pieces in the package. There were twenty-seven.

"Oh, by the way, he knocked a splinter off the corner of the cabinet, so it will be in there too. But we could not decide what were violin pieces and what was the cabinet corner, and so my wife thought I should bring it all."

Anton took a deep breath, and stirred the fractured pieces of wood. He could identify the scroll and the neck of a violin, but the other pieces were so shattered they were unrecognizable as being part of a violin.

Mr. Chou mistook Anton's silence for refusal and became anxious. He began pleading. "Please, please, can you save this instrument? I realize it will never be able to make music, but if you could just restore it to a violin shape I would be very, very grateful. You see, this violin was a gift to me from my grandfather. It was his violin. I have always treasured it."

Anton slowly picked up the motley pieces and laid them into a jigsaw of a violin shape.

"Please, I don't care what the cost is, just as long as you can get it back at least into the shape of a violin for me," pleaded Mr. Chou. "I want something to remember my grandfather by."

"Yes, I can do it," Anton muttered to himself. "Of course I will try help you," he said aloud, looking from the pieces to his visitor. "Just give me time."

"You can have all the time you like!" exclaimed Mr. Chou.

Eight weeks later Anton rang the Chou home.

"This is Sie Anton. How's your boy's *kung-fu*?" he asked mischievously when Chou HuiRen answered the phone.

There was a surprised pause, and then: "It's banned!" declared the boy's father emphatically. "Have you been able to do anything with those bits of wood?"

"Yes, you can come and get your violin," said Anton.

"Oh, thanks very much. I thought you would take longer to get it fixed. I really appreciate your help. I'm pretty busy at the moment, but I can come over in three days' time to pick it up."

Anton realized that Mr. Chou was expecting only a violin-shaped *objet d'art*, so he answered, "Sure, come whenever you are ready. You will be very happy with the way it looks!"

Three days later Mr. Chou was indeed very happy with the look of his restored violin.

"Wow, that's amazing!" he exclaimed. "I can't even see where you have mended it! It looks as good as new!"

"Would you like to hear it?" asked Anton, his eyes twinkling.

"Hear it? You mean it still works? Someone can still play it?" asked the incredulous Mr. Chou.

Anton lifted the violin to his chin, picked up his bow and began to play. The beautiful sounds of Jenő Hubay's *Violin Makers of Cremona* filled the room. When he finished a pregnant silence followed. Finally, it was broken by Mr. Chou's gingerly taking the instrument from Anton, and looking at it in amazement.

Chou picked up the bow and began to play. Joy radiated from his face.

"Why, it sounds better than before!" he exclaimed incredulously, lowering the violin and gazing at his beautifully polished instrument. "This violin was such a precious family heirloom. I can only say that now I'll not only remember grandfather. I will remember you forever!"

"Just make sure you keep either your son or the television locked up!" laughed Anton.

"Or the violin!" quipped Chou HuiRen.

*****

Although the Baptist College had refused Anton's application to lecture, during the visit of Carleen Hutchins Anton met some lecturers from ChungChi College in Shatin.

Dr. Ryker contacted Anton and asked if he was willing to help a student, Thomas Yuan, who was writing a thesis on violin making. Anton joyfully accepted the position of temporary tutor at ChungChi College. He enjoyed working with the enthusiastic young student, who did very well with his thesis and made an exhibition of his work at the College.

In appreciation for Anton's help Thomas Yuan placed a photo of Anton at the top of his poster-style exhibition. After years of stress and discrimination Anton was grateful of this honor. But unfortunately this recognition caused him serious problems.

Two other violin repairers in Hong Kong, whom Anton did not personally know, resented the fact that "unknown" Anton was asked to help the student's research. They made threatening phone calls. Although Anton's Cantonese was very poor, he understood he was being called a "rascal" and there were threats to burn his house.

Once a stranger came to him in the street and began saying similar things. Anton could not understand all he said, so he simply repeated in Mandarin many times, "Can't you see I am a poor worker?" Unfortunately, this distressing harassment continued for a couple of years.

*****

1984 was a doubly significant year for Anton. First he applied for, and at long last received, permanent Hong Kong identity cards for himself and YuXiang. With these he was able to obtain a British National Overseas Passport, and finally visit his family in Indonesia. 1984 was also the year he entered the Hong Kong real estate market.

The Indonesian visit was was not easy, and unexpectedly most painful. Although Anton obtained passports as soon as he legally could, before he could return to Indonesia he learned his father had also died. The basic reason for leaving China had gone. But he looked forward to seeing the familiar sights, meeting as many people from his youth as he possibly could, and especially catching up with his brother and sister whom he had not seen for almost thirty years.

It was no surprise that Indonesia, especially its cities, had modernized, but in many ways life in Kudus was much as he remembered it. First he went alone to the family home, an occasion he knew would be sad one. He walked around the old home remembering scenes from childhood, handling a few familiar things that still lay in the house. Complex emotions flooded his mind. No one currently lived there because his parents' estate had not been settled. As he liked to do at significant times in life, Anton took a video of the occasion, capturing himself pumping water from the old well in the back garden with the well-remembered banana trees. Inside the house chickens scurried around, just as they had done when Japanese soldiers had intruded and made his mother cut her hair and blacken her face. The floor was covered in chicken dung. The chickens were skinny, long-legged birds that also gave Anton memory flashbacks to watching cockfights with Budi. A stray kitten chased a small ball in the old bedroom, unmindful of the messy chickens and the sad emptiness of the old home.

Thinking he had dealt with the tough emotional part of his visit, Anton walked across to visit his brother and sister-in-law. But what a shock! His brother's letters always sounded cheerful, full of chatty news, and Anton believed he was content toiling in the textile factory assigned him by the government. He was happily married and never referred to his unfulfilled dreams.

Nothing therefore had prepared Anton for the shock of seeing his brother's real situation. No one met him when he called from the gate, which surprised him. As he walked through KiemSiang's front door he saw his brother lying on a bed, a hollow-eyed, shriveled-skeleton of a man. Hacking coughs wracked his yellow, wasted body as he attempted to get up and greet his long-absent brother.

"KiemGiok! KiemGiok! How good to see you!"

With difficulty Anton tried to act normally. He talked about trivial details of his trip, and made platitudinous comments about obvious changes he had noted in Kudus. His brother smiled happily but said little.

"How long has he been like that?" he demanded of his sister-in-law as soon as he could privately.

"About three months. I . . . I couldn't bring myself to tell you! I just hoped you could get here before . . . before he died. I . . . I am so glad you are here! We've been to many doctors, so many doctors, and many hospitals, but in the end they all told us the same thing: there is no treatment, and nothing anyone could do."

"Does he have cancer?" asked Anton, emotion making his voice sharp and rough.

"Yes," replied his sister-in-law, turning away to hide her tears. "Honestly, I really tried to get someone to make him well. When I noticed he looked sick, I tried to get him to see a doctor, but he always insisted he was just fine. Till it was too late," she ended with a sob.

"Of course," answered Anton, gently. "Of course, I understand. We just have to make sure he is comfortable now."

"Yes, the local doctor is doing that. He is a kind man."

KiemSiang always acted his old cheerful self, deeply interested in Anton's experiences in China and Hong Kong, but he could say little without being wracked with coughing. Anton, not normally talkative, did his best to describe details from his life for his brother. KiemSiang clearly enjoyed listening to his brother's adventures. But it was distressing to watch him try to walk. It was so hard to believe, so hard to understand. KiemSiang was only forty-four.

Because of his brother's sickness, Anton did not talk to him about his experiences in Bandung. But little by little his sister-in-law shared what had happened. KiemSiang was a top student, and his sunny disposition endeared him to everyone, staff and students alike. But after an attempted coup on 30th of September 1965, everyone became afraid of everyone else. The government blamed "communists", but no one knew who they were, except that China was a Communist country. Like many Chinese KiemSiang was taken for questioning. There were many people to vouch for his loyalty, but no one trusted anyone. After questioning he was expelled from the university, and later assigned to work in the textile factory. He was very grateful to return to Kudus.

"Our parents did not dare tell you."

Anton shook his head slowly. "While we were having a tough time in China, with little to eat, I had no idea my family was struggling. Sometimes when things were really bad I used to dream of the wonderful days in Indonesia. How little did I understand!"

"Oh, the horrible times did not last long, and we always had plenty of good food," sister-in-law smiled. "And now, it is really good here. Indonesia is a good country now!"

Anton looked around his brother's simple home, and suddenly felt enormously rich. He was grateful for the changes in Indonesia, changes that allowed him to visit his homeland. He realized every country has bad times, and it is better to focus on the good. But he had to admit that the standard of living he now enjoyed in Hong Kong was significantly higher than that of his brother and sister.

Although Anton made extensive enquiries he found nothing about his beloved guitar teacher Mas Toha. It was thirty years since they had worked together, thirty years when they could not communicate because Mas Toha was virtually illiterate. Anton hoped that somehow he would find the man who had shown him music could be made in any situation by anyone. He was also unable to find his old friend Budi. The *kretek* factory had gone, and although he walked out to Budi's old family home, there were strangers living there. He made extensive enquiries about Anton Piontek, but they all drew a blank. No one could tell him what had happened to him, or where he had gone.

When his visit ended, it seemed his trip had accomplished very little, little that he had dreamed of. As Anton prepared to leave Indonesia he felt frustrated and sad. He had hoped at least to finalize his parent's estate, but because of his brother's health, he dared not broach the subject. KiemLiang, his sister, showed him their parent's will, but given the difficult situation, he decided to do nothing without discussing the issues with his Hong Kong family.

A few months after Anton's visit, KiemSiang quietly passed away.

Anton was relieved that YuXiang and both his daughters agreed that his parents' estate should be divided between his sister and sister-in-law.

"After all," said YuXiang generously, "we are doing fine here, and from what you say they could do with help."

In 1988, as soon as he could arrange it after his brother's death, Anton returned to Indonesia and signed all necessary documents that gave his two sisters, by blood and marriage, legal ownership of the family home. He felt great peace when it was all settled, confident that he had done his best for his Indonesian family.

On this second trip Anton had time to visit his old school in Semarang. Whilst there were no teachers who remembered him, many were interested to learn about one of their former pupils. Much to Anton's surprise, the teachers organized a for reporter from the magazine *Intisari* to interview him. Anton was amazed at the interest of the reporter, and his fascination that a locally born person was involved in specialized violin construction and repair work. His sister-in-law promised to send him a copy of the paper when the article appeared. She was true to her word. This magazine article had a delightful sequel: a few years later Anton Piontek saw it, and through it made contact with Anton.

Three years later his sister-in-law wrote to say his sister KiemLiang had died suddenly but peacefully. She was not even sixty years old. It seemed all Anton's ties with Indonesia were being cut.

*****

By mid 1983, as Anton looked back on his life, the misery of the early years in Hong Kong was being erased by solid accomplishment. The support of Jan Van den Berg, the opportunity of working with Carl Pini for the Don

Pascuale opera, the Hutchins tour, the dramatic improvement in YuXiang's health, NeeZi's scholarship to study at the Vienna Conservatory and YanYan's plan to study in Sydney, Australia, the Chungchi College exhibition, and of course his increasing reputation as a repair luthier were all encouraging. He began to develop a most daring plan: to buy his own home. Hong Kong rents were extremely high: about seventy per cent of his total earnings went into rent. Although the Sie family lived most frugally, this high rent meant they could never improve their lives unless they owned their own home.

In September 1982, just after his own visit to China with the Hutchins, Margaret Thatcher, Prime Minister of the United Kingdom, also visited China and began negotiations with Deng Xiaoping to return Hong Kong to Chinese rule. Everyone in Hong Kong was aghast and agog with the astonishing idea that Britain would freely give its entire colony back to China when the New Territories lease expired in 1997. Anton remembered his idle thoughts about what would happen to Hong Kong in 1997 when he had arrived in Hong Kong twenty-seven years earlier. Britain, of course, correctly recognized that without the New Territories the rest of Hong Kong was not a physically or economically viable entity. However, many Hong Kong residents had escaped from China when the political situation there became difficult, and they had no love for the Chinese government. They had no desire to live under communist rule. The Sino-British talks suddenly meant there was a huge demand to immigrate to other countries. The United States of America (called *Mei Gwok*, or "the beautiful land" in Cantonese) was the most favored destination, closely followed by Britain and Canada. Australia and New Zealand were less known, but quickly became popular as people jostled to leave the colony before they considered it would be too late.

This had a dramatic effect on the economy of Hong Kong generally, and its housing sector in particular. As newspaper headlines continued to scream the drastic news of an imminent return to China housing prices fell steadily as the properties of fleeing people flooded the market. Anton decided to watch developments and look for a chance. Unlike other people in Hong Kong he did not fear China, or its government. He was appreciative of both the education and opportunities China had given him.

"All this bad press about Sino-British government meetings is makings people very frightened and hopeless," he announced expansively to YuXiang and YanYan one evening.

They looked at him blankly. "So?" returned YanYan saucily, when Anton said nothing more. "Everyone knows that!"

"China will take Hong Kong back. I know it will happen."

The two women just looked at him. Why was he stating what everyone in Hong Kong, indeed the whole world, knew?

"Housing prices have dropped dramatically as people try to sell their homes, and no one wants to stay in Hong Kong beyond 1997."

More silence, just quizzical looks. Had their father and husband only just got around to reading the news? This was embarrassing!

There was a long pause, then Anton cleared his throat and declared dramatically, "So, now is the time that we should buy our own house!"

"Anton!" exclaimed YuXiang, horrified. "How could we?" A lifetime in communist China had expunged all ideas of private real estate ownership from her mind.

"Hey! Dad! You're right!" grinned YanYan, gazing at him in admiration. "So that's what you were getting at with all those dumb comments? I'll help you! I certainly will!"

It was Anton's turn to be silent. He looked at his daughter, only fifteen. Surely, she was far too young to know anything about property and houses and how to buy them. But she did have one gift: she could speak Mandarin, Cantonese, and English fluently. She was doing well with her studies, and probably did have time to help with this project.

"Yes," said Anton, "it would be very good to have your help." He looked at his wife still shaking her head incredulously. "We will work together on this project. It will be *our* house." He stressed the "our".

Never had so many newspapers, Chinese and English, arrived at the Sie household. Never did Anton take less interest in YanYan's violin practice. Never did YanYan race through her homework so fast. Never did YuXiang take so much interest in the property market. Never so often did Anton add and subtract, subtract and add, and then once more recalculate his finances. For more than a year the family carefully studied the real estate market and their financial resources.

Then in the summer school holidays of 1984 YanYan, with her parents' support, began contacting real estate agents. Perhaps it was fortunate that it is notoriously difficult to pinpoint the age of Chinese women. What would they have thought if they had known they were dealing with a mere teenage schoolgirl? She spent a great deal of time looking at apartments. Even when school restarted for the year she continued her energetic housing search after classes. Both Anton and YuXiang knew their lack of Cantonese was a serious handicap in this house search, prejudicing their chances of getting a fair deal from agents, and they were content to allow their young daughter to do the donkeywork, and give them daily progress reports.

Towards the end of the year YanYan found an advertisement for an apartment in the highly respected Mid-levels area of Hong Kong Island.

YuXiang and Anton agreed that it looked very suitable. YanYan inspected the apartment with the agent and she was impressed.

"Father, this place is so big! It isn't modern, but it's very comfortable!"

Anton looked at the advertisement. Of course YanYan would think it was large, he thought. At twelve hundred square feet it was almost three times the size of the place where they were living. But her enthusiasm was so great, and the price seemed so reasonable, that Anton decided the time had finally come for him to be personally involved. Two days later he accompanied his daughter and the agent to inspect the apartment. He made sure all discussion was in English.

Anton's appreciation for the apartment was love at first sight. Yes, the building was old style, but the eighth-floor apartment was large and commodious. One of its especially attractive features was a long semicircular balcony with broad views of a leafy green hillside and Hong Kong Harbor, a place that would be perfect for his steadily increasing collection of birds. However, Anton was puzzled that the apartment seemed at least four times the size of the apartment they were renting. When he enquired further he discovered the agent had taken him and YanYan to the wrong apartment, not the one they had originally seen advertised, and priced! More seriously, this apartment was not only larger, but also a little more expensive. Anton, however, decided to buy.

Unfortunately, he had not reckoned on YuXiang's vigorous opposition. Although she had been enthusiastic about their move to Hong Kong from China, she still had a lifetime of communist anti-capitalistic indoctrination to overcome.

"It's ridiculous!" she snorted, when she heard the price. "Utterly ridiculous! You admit yourself that we don't have even a third that money. How can we possibly buy such a place? A small apartment would be more than adequate for us!"

Anton pled, explained, re-explained and pled some more, but YuXiang was adamant. She refused to even go and look at this preposterously expensive house.

Anton, however, was sure the place was right for them. Quietly he visited his old friend Tseng HsiangHo, the kindly musician who had been guarantor for the rental contract of the Queens Road apartment. Perhaps he would know a way around the impasse. He poured out his longing for security and a home of his own.

"So YuXiang is against this apartment is she? I guess real estate was not exactly something we got into back there in China!" HsiangHo laughed. "You can't blame her for being cautious."

"True. But rent is just eating up my money, with nothing to show for it. I don't think I will ever get a better opportunity. This place is all I could ever want."

Tseng HsiangHo stroked his chin thoughtfully. "You're right. Already there is talk that the agreement with China won't be too bad, and people's confidence in Hong Kong will start to return."

"That's exactly what I say. But I have to admit, the deposit on this place is more than I was prepared for. I told you, didn't I, that we had originally planned to buy another place?"

"Yes, yes. I see your problem. Would it help if I loaned you the difference in deposit? That might make YuXiang feel more at ease."

"I'm not sure, but I think the deposit may be her big concern. The problem is she just won't talk about it. You see, we did all our repayment calculations on the cost of the bank loan for the first house. But I promise you, I will repay you even before the bank!"

"You better not do that!" laughed HsiangHo. "Anyway, tell YuXiang I'm happy to help and see if that makes her happy!"

Perhaps it was Tseng HsiangHo's generous offer. Perhaps YuXiang had been making her own enquiries and calculations. Perhaps she just got tired of fighting. Whatever the reason, after two days of vigorously opposing the idea, she finally agreed to visit the apartment. Like Anton, it was love at first sight. What Anton did not appreciate was YuXiang realized that for the first time in their married lives this apartment offered a separate workroom for Anton. At last, at long, long last, she could contain, at least to some degree, the constant mess of wood shavings and dust from his violin making, mess that had distressed her for so many years.

Thus, late in 1984 Sie Anton and his wife Wang YuXiang became the proud owners of their first home. The apartment was old and required a lot of redecoration, but Anton was happy to do this himself. Little by little he painted and repaired and each time he went to his new home he took some of their possessions from the Queens Road flat. They finally moved into their very own apartment on the 28th December. Bank repayments were steep, many times what they had lived on for their first years in Hong Kong. But they were used to living extremely frugally. By the time Hong Kong returned to China they were debt free. Anton admits he worked incredibly hard during those years, at least ten hours a day. He accepted any work that came his way. But he was convinced that the God of eternity had led him to inspect the wrong apartment, and that securing this home was his special good luck. He was extremely grateful that he now not only had many students, especially at the Tsuen Wan Adventist Hospital, but suddenly many orders for making violins. At last he not only had work, but a place he could call his own.

*****

Soon after they moved into their new home, because of the support and recognition of Carleen Hutchins, Anton was contacted by two international publications. Swiss luthier Claude Lebet included Anton in Volume 3 of his 1985 *Le Dictionnaire Universel des Luthiers*. Anton was also included in the revision of the comprehensive reference book on violins and luthiers *Die Geigen and Lautenmacher von Mittelalter bis zur Gegenwart* (The Violin and Lute Makers from the Middle Ages to the Present), last revised in 1922. Since these books were updated only about once every fifty years, Anton recognized that his inclusion in these works was another piece of amazingly good blessing from "eternity".

Anton's reputation was expanding internationally.

*****

Buying their own home meant celebration time again, but this time Anton decided it would be a grand event. He would invite all his students and their parents to his home for a party.

Anton was never a particularly social person. Although YuXiang was quiet, she enjoyed mixing with her friends. As their circle of acquaintances increased both of them were frequently invited to "old China friends" social occasions. YuXiang loved these times when she could chat about China and freely speak in Mandarin, but Anton rarely accompanied her. At first it was because he was working very hard and totally focused on feeding the family, educating his daughters, and repaying house loans. But if he had been truly honest with himself and YuXiang, he would have admitted he did not enjoy small talk; sitting around sipping tea and cracking pumpkin seeds bored him to tears. Although he enjoyed discussing serious topics with friends, he did not appreciate having to wear uncomfortably fine clothes to discuss the weather and price of handmade suits. He even avoided parties given by close friends.

But inviting his students would combine work, celebration and party. They could play their music together, which he would supervise, and he would provide Indonesian food for them to enjoy.

It was not the first time he had invited a student to visit his home. Soon after his visit to China with the Hutchins, the author, mother of his students

Sven and Genevieve, had asked, "Where did you go for your holiday?" Well, it was not exactly a holiday, Anton admitted. When he explained his trip to China and his intention to craft fine violins, Elizabeth showed great interest, somewhat to his surprise. She even suggested she would like to see his violin-making work.

Anton thought about this, and finally decided that Sven and Genevieve probably would benefit from learning about violin making. So they were invited to the Queen's Road apartment. YuXiang shyly kept herself hidden in the bedroom, but Anton showed them the basics of violin construction. He later admitted he thoroughly enjoyed this social occasion. Sven was delighted to be photographed holding an expensive piece of Norwegian spruce.

"Fancy a small chunk of old wood costs thousands of dollars!" exclaimed his mother.

The Ostring family remained interested in Anton's work, and talked about it to others in their apartment complex. Soon everyone at Tsuen Wan Adventist Hospital knew Anton both taught violin performance and made violins. Lui Oliva and Sven Östring were keen young violin performers. Anton presented them for examination by the British Royal Schools of Music, and both did well. Soon there were six more violin students, and a mother learning guitar at Tsuen Wan. Lui's mother, Mae Oliva, a keen amateur musician, started a small children's orchestra, and somehow thrived on the ghastly noises the children initially made. While Anton's Tsuen Wan days were busy with serious learning, Mae Oliva's friendly efforts eventually produced a promising youth orchestra.

Anton discussed the party celebration idea with YuXiang. He explained that the people at the Adventist hospital were mostly vegetarian, and therefore easy to cater for; chicken could be offered to his other students and their families. He planned to make Indonesian food, which he was confident everyone would enjoy. The mother of another of his Tsuen Wan Adventist Hospital students, Caroline Catton, often asked about Indonesian cuisine.

"Yes, it's a good idea," YuXiang responded, happy he had offered to do most of the cooking. "I think YanYan could help us. Her English is good. It won't be long before she goes to Australia to do that course on food technology, so it will be good for her to practice English. She could play violin with your students, and help me with food and conversation."

"I'm glad you agree," said Anton, grateful of her support.

"But we can only do this if and when the house is completely redecorated," declared YuXiang with a sly smile. "It's a lovely home, and I want it to look its best!"

Anton was about to make a speech about how hard he worked, when suddenly he grinned. "OK. I'll get all those little maintenance jobs finished," he smiled.

But getting everything organized took longer, much longer, than expected. Anton kept his promise and completed all the redecorating projects, but it took time.

Finally, at Chinese New Year, in January 1986, the special occasion arrived. By then YanYan had unfortunately left for her studies in Australia, but fortuitously NeeZi was was home from Austria. She was more than happy to assist. The party was a resounding success, in every way.

After her strenuous study program in Vienna NeeZi found it relaxing to be the "experienced violinist" with this group of young players, and enjoyed herself. With NeeZi nearby and willing to do necessary translation, YuXiang also enjoyed herself, happy to circulate amongst her guests, smiling graciously. The students were delighted to show off their ability, and performed cheerfully. And as for Anton, not only was he the undisputed leader and director of everything, but his Indonesian cuisine was an outstanding success, with the mothers of the Adventist Hospital students busily noting recipes.

As a special, joyous bonus, the students made friendly acquaintance with Anton's pet birds. Mr. African Grey was the star, as he chatted cheekily to all who passed by!

"Isn't it wonderful!" exclaimed Anton as the last of his guests left. "Today we were many people from many different countries, yet we all enjoyed each other because of music!"

YuXiang smiled. "Yes, we did. Those people from that hospital are certainly very friendly."

"Yes, they've always been like that. But it's more than friendliness. They make you feel like a real person. I wonder if it has something to do with their religion?"

Anton paused. "Anyway, I think we all had such a good time we should have another student party."

They did, three years later. It too was a great success.

*****

Despite what a few "opposition" violin repairers had said and done, Anton's repair work flourished, and he now had many clients. More and more international musicians asked for his help. He met South African Vincent Fritelli with his lovely Guanarius violin, David Arben from the Philadelphia

Orchestra, and Edward Greisman with a beautiful Vuillaume cello. Anton was particularly delighted with the opportunity to repair two violins owned by the Swiss consul in Hong Kong. One was a Stradivarius, and the other a Guarnerius played by Blaise Calame.

Even more exciting, Anton began to get orders for his own violins. Ngai HoChao, who had been a member of the Hong Kong Philharmonic, and a Mr. Inagaki both ordered his instruments.

One day late in 1985 Anton was very surprised to receive a phone call from a reporter working for the English language *South China Morning Post* newspaper. He politely asked if he could interview Anton in order to write a feature article for his newspaper. Anton was guarded: how had this reporter heard about him? He remembered all too well the trouble that helping Thomas Yuan at ChungChi College had caused. He was not anxious to repeat that ordeal. The reporter was evasive about the source of his information, but persistent in his attempts to gain Anton's permission. Finally, Anton agreed to the interview.

When the reporter, Vernon Ram (with a photographer), came to his home Anton insisted he was primarily a scientist. To his surprise this made the reporter become even more interested.

"The English readers of our paper are highly discerning professional people, and they will be very interested in your story," he stated. "Do you mind if I take photos of you working?"

"Not at all, working is what I like to do," responded Anton, smiling. He quickly moved into his workroom, and picked up his tools. But the reporter remained in the living room.

"These are interesting pictures you have," Mr. Ram commented, surveying the artwork hanging on the walls. "Where did you find such a good picture of Stradivarius? I don't think I have ever seen this one before."

"Oh, those," said Anton, emerging from his workroom.

When Anton explained he had painted the pictures himself, the reporter's interest increased even more. He now wanted to organize photos of Anton working with the picture of Stradivarius in the background. Ruefully remembering his painful experience in the Ocean Painting Studio, Anton cautiously agreed. From being an ignorant Ah Chan earning a mere thirty dollars a painting he was suddenly, eight years later, to become the focus of the cultured English-reading sector of Hong Kong society, pictured working under one of his own paintings!

The double-page article duly went to press, and received considerable and favorable notice. What Anton did not expect was that the English article triggered several articles about him in the local Chinese language papers. Anton received many phone calls and invitations about his work, but he politely yet firmly refused further contact with the callers. Unfortunately, as he feared, some calls were not friendly, and he felt compelled to terminate one of his phone numbers.

However, there was one invitation he just could not resist. The Hong Kong City Polytechnic invited him to apply for work in their physical laboratory, specifically to work with their superconductor project. They even sent an official application form. Anton could not resist filling it out and sending it off. But in the end he was (as expected) rejected because, again, he did not have British qualifications. Despite recognition in the Hong Kong press, he was back to where he had been eight years before, rejected by the Baptist College and Hong Kong University as a science teacher. But now, it did not matter. This time Anton could smile. The Acoustical Society accepted him as a full member, and *Who's Who* accepted him. He was recognized internationally, even if he could not be accepted as a scientist in Hong Kong because he lacked magical British qualifications.

*****

Most days when Anton returned from teaching he went straight to his workroom and continued working on another repair assignment, or more excitingly, crafted one of his own instruments. He would remain happily buried in his workroom until YuXiang called him to eat, and he reluctantly stopped work. The apartment was very quiet now, with both daughters overseas studying. NeeZi was enjoying her time at the Vienna Conservatory of Music, and doing well. Anton did not quite understand YanYan's interest in food technology, but it was a form of science and therefore he was supportive. He was delighted when she wrote saying she had joined the university orchestra, and was enjoying the experience. So, she loved music too!

YuXiang missed her daughters very much, but since her improvement in health, from Anton's point of view she seemed very content with life. He could not remember when he had last seen her lying around reading a foolish story as she had done for months, years, prior to her surgery at the Ruttonjee Hospital. She did not even ask for extra money as she had sometimes done before. Not that she was extravagant or demanding, but he could not help noticing that she seemed content to live completely within their frugal budget. The payments to the bank were ahead of schedule, and Anton looked forward to the great day when they would be debt free.

One day he arrived home earlier than usual and was surprised to hear a murmur of voices. Who could it be? Neither of the girls was due home on vacation for many weeks. What a pity, he often thought, that the girls studied in different hemispheres! NeeZi's holidays coincided with Hong Kong's summer, but YanYan's vacations from Australia were at Western and Chinese New Year.

While he was thus pondering as he walked towards his workroom, YuXiang came out of the girls' bedroom accompanied by a young woman. He knew YuXiang had set up the room as a study for herself, but who was this young woman?

The young girl bowed her head politely towards Anton, and hesitatingly said, "*Nihao*" (hello in Mandarin). YuXiang beamed approval, and carefully said in slow, stilted Mandarin, "Come again next week."

"Who's she?" asked Anton, when the door closed after the girl.

"Oh, she's one of my students," answered YuXiang with a mischievous twinkle in her eye.

"Students? What are you teaching?" asked the surprised Anton.

"Mandarin."

"Mandarin? Really? How long have you been doing that?"

"Oh, since moving to this apartment. I've met many people who think it would be good to learn Mandarin because Hong Kong is going back to China. I think most want it for business purposes, but some just to learn."

"How long have you been doing this?" Anton repeated.

"Oh, a year, maybe more than a year, ever since we came here."

"How did you get your students?"

"It started with just one young woman. Her mother met me at the market and noticed my Cantonese was poor. 'Do you speak Mandarin?' she asked, and when I nodded yes, she asked if I would teach her daughter who had just got a job with a company that traded in Shenzhen."

YuXiang paused. "You know that little bridge we walked across with your two aluminum boxes? Well, it seems there's quite a town there now!"

Anton looked at his wife. He was about to say something like "Everyone knows that!" when he remembered she had little opportunity to travel. Instead he said, "So you got one pupil?"

"Yes, then she told her friends, and friends told their friends and now I am really quite busy. When I am not teaching I spend a lot of time preparing lessons. See here," YuXiang went to a drawer and pulled out a thick file. "These are what I use to teach. I have lessons for young children at school, for young adults working, and some for older adults who want to be ready when China takes over."

Anton stood there, speechless with approval and surprise. One look at YuXiang's file and he realized she was a meticulous and obviously capable teacher.

YuXiang smiled mischievously again. "Remember how you said this would be *our* house? Well, now I am doing my part so it really is *ours*!"

The triumph and pride in her voice was palpable.

"You are very good," said Anton, smiling. But expressing appreciation was not something he found easy.

"You know," began YuXiang, and then looked down at the floor shyly. After a few seconds she looked up, pride radiating from her eyes. "You know, we could live off the money I earn now!"

"We could what!" exclaimed Anton. There was a stunned silence.

"Yes, even the bank payments."

Anton had always recognized his wife as beautiful, kind, and hard working. But the daily bus-trip reminders in China of her inferior educational status obscured his recognition of her true potential. Suddenly he realized she had the ability to teach, a capability that truly matched his own. He gazed at her, as though seeing for the first time her full value. She had sometimes said she would like to teach, but with no training he considered this a mere pipe dream. Now she was achieving her dreams just as he was his!

"When would you like to eat?" asked YuXiang, breaking the silence. She walked across to the bench. "Oh, by the way, I thought I could save up and both of us visit NeeZi in Vienna. Remember, she is graduating soon," she added.

Together they laughed with joy.

*****

Anton's confidence in himself and life was growing. He worked extremely hard, and by the end of each day was often totally exhausted, but after more than a dozen years in Hong Kong he could hold his head high. Some had opposed him, even bitterly, but generally he felt accepted, a respected citizen of the colony.

Convinced he had found really important information about violin plate tuning, he maintained correspondence with Carleen Hutchins on the issue. As he repaired some expensive and beautiful violins he daily appreciated their magnificence. But he became more and more confident that their value did not lie in some mystical inspiration available only to the ancient luthiers, now lost in the mists of time. They were wonderful instruments because their makers had obeyed the laws of science, and with care and precision crafted them to conform to those laws.

In 1988 Anton received significant confirmation of his violin acoustics theory. He was able to make a violin whose plates had three matching *Eigen* frequencies in octave relationship. This was the only violin with such characteristics in the history of the Catgut Association Society, and Anton was extremely proud of it. It had a beautiful, loud, and clear tone, and was tested by several violinists who appreciated its worth. He sold it for a large sum of money, but was sorry to part with it. Following this break through he made three more tri-harmonic matching violins, but it was very difficult work and Anton could not consistently get the three harmonics to match. However, bi-harmonic violins, which he regularly made, also have very good tones.

Despite being a dedicated scientist, Anton enjoyed having fun in his work. He made a violin in the style of a Chinese dragon. This was a daring concept, as a dragonhead does not easily fit the basic neck scroll of a violin. But after three attempts he finally made one that satisfied him. He painted dragons on the back and front plates of this violin, choosing classic pictures from historical times in China (including a design that graced the gown of Empress Cixi). Even the *f*-holes were dragon designs. He was meticulous in its plate construction, so its sound was both powerful and beautiful. The violin was finished in time for year of the dragon, 1988, a fitting memorial of Kowloon, now home to him (Kowloon means Nine Dragons). There were, appropriately, nine dragons on this violin. It received much favorable comment when Anton entered it in the Hong Kong Art Centre Exhibition, 25 October – 17 November, 1990.

Another "fun" enterprise was the construction of a five-stringed violin. Schubert had written a sonata called the *Arpeggione*, and its five-string harmony was usually played by a cello and viola combination. Anton decided it would be interesting to make a single instrument that had the required five strings, and he called his invention the *Pentaline* violin, *penta*, of course, meaning five.

Over time Anton and Carleen Hutchins exchanged hundreds of letters. He offered to make an "openable" violin (one that could be taken apart easily) and send it to Hutchins' research laboratory for examination. The one he sent was his sixty-seventh violin. He made it from Chinese woods, demonstrating that although good wood is important, it is one factor among several, and many woods if properly prepared can make a good violin. The Americans researchers were very impressed. They rated this violin as equal in its acoustic qualities to any Stradivarius or Guarneri violin. Anton observes that the tone of a violin is like taste in eating, some like one thing, and some another, like the Chinese proverb, "If you love her, then she is the most beautiful."

Carleen Hutchins, true to her open minded approach, studied Anton's violin and opened some of her own. Eventually she agreed that he was right about violin plate tuning, and decided, in honor of his work, to use his own term for the method he had developed, the Anton Sie Bitri plate tuning. Anton was not quite satisfied when she first called his method the "Bitri Octave" plate tuning, but at least the idea was being presented to the world. True to her commitment to science, Carleen Hutchins published his findings in 1991, and agreed the method of plate tuning Anton had perfected was significant, and it should be named "Bitri harmonic matching plate tuning" in his honor. It was a triumphant moment for Anton.

Seven years later, in 1998, she wrote fully acknowledging his contribution to the understanding of violin acoustics and plate tuning. Anton was extremely pleased that his method was now recognized worldwide and available for anyone who was interested. Like Carleen Hutchins he was not interested in personal fame or recognition, but simply that true science be recognized.

Although Americans were now convinced of the value of his findings, he received no feedback regarding his method from China, Hong Kong or Europe. He later discovered that there were many errors in the published translation of the lectures given by Mrs. Hutchins, which hampered acceptance in China. Unfortunately, Europe still defensively believed that the art of the Cremona luthiers was both unbeatable and lost, despite Anton's presentations, and Mrs. Hutchins' endorsement. Anton is convinced that it is not "inspiration" that is needed but dedication to the science of acoustics and meticulous attention to precise workmanship.

In the USA Anton received several awards. Carleen Hutchins had not only asked him to be a member of the Cat Gut Acoustical Society, but she also recommended him for the 1990 publication of *Who's Who*. Anton jokes that he was simply lucky to be included in this publication. However, Hutchins also recommended him for membership in the Physical Society. After careful investigation of his research Anton was accepted as a full member. He is very grateful to Hutchins for giving him these opportunities for recognition.

A few of Anton's friends thought he should get more recognition for his research. But he is not concerned. He is glad his research has finally been acknowledged. Carleen Hutchins, he says, was a great scientist, willing to recognize that her original findings were not right, and correct them in the light of his study. She deserved all the honor she received. He believes she was an unselfish person, indicated by her willingness to recommend him for honor. He calls her a "true friend, a beloved teacher, and collaborator." As he had so often thought, in music and science the baton must be passed on.

Life was again changing for Anton. In 1990 his students Lui and Oyen Oliva emigrated to Canada, and a year later the Ostring family moved to New Zealand. He missed these friendly students, but was no longer dependent on teaching for his livelihood; anyway, many students now wanted to learn from him. His daughters were well educated, and a few years later the Mid-levels apartment was paid for. More significantly, his own violins were now in demand.

Looming on the horizon was the momentous transfer of Hong Kong's sovereignty from Britain to China. Anton looked forward to this with happy anticipation. He decided to celebrate the occasion by making another special violin. It would not only be a beautiful sounding instrument, but he would give it a special Chinese design. He planned the instrument for years, and finally decided to use the idea of the supernatural *Devi* legends, from pictures in the ancient *Dunhuang* Caves. The *Devi* figure was usually depicted playing a *pipa*. On a visit to the Shangrila Hotel in Shenzhen, across the border from Hong Kong, he found a classic bronze statue of the *Devi* figure, exactly the image he wanted. His sketch transformed it into a classic Chinese painting, which he then portrayed on the back of the violin. For the neck of this instrument he crafted a Tang Dynasty figure of a woman. The resulting violin was beautiful to look at with exceptional sound. All Anton's violins are named, and this one was called *Souvenier de Hongkong*, a memorial to Hong Kong's reunion with the mainland of China.

When all the pageantry and fanfare of the transfer of sovereignty from Britain to China finally occurred in 1997 Anton and YuXiang watched proceedings on their television with great interest. They had no apprehension or concern, and felt a deep sense of pleasure that they were now once again part of the country that had been good to them in their youth. They looked forward to the future with confidence.

*****

With the death of his sister, Anton thought connections with Indonesia had been severed. Then one day early in 1999 he unexpectedly received a letter from his beloved violin teacher.

"Look at this! Look at this!" he exclaimed excitedly to YuXiang, fluttering the letter in front of her. "It's from Anton Piontek!"

"Who?" she said absent-mindedly, as she counted the patterns in her knitting.

Anton stared in horror. Surely she knew who Anton Piontek was! But then he remembered, she probably did not.

"He's special . . . he's my violin teacher! And I've found him!"

"You mean you lost him?"

Anton felt more frustrated, so he simply explained how he had tried to find his teacher when he returned to Indonesia, but without success. Now Anton Piontek had written saying he had seen the article about Anton in the Indonesian magazine *Intisari*, and by writing to the magazine had discovered Anton's address.

"Wow, that's a story!" declared YuXiang, at last suitably impressed by the letter.

"You know, without him I'm not sure where I'd be today. Not only did he teach me violin performance, but he also taught me the basics of violin repair. So really he's the one who started me on the road to be a luthier."

"You owe him a lot," said YuXiang.

Anton sat down and reread his letter. "It sounds as though his health is not good, and he mentions that his violin has not been performing well. He even wonders if I can repair it."

"You mean you must go to Indonesia again?" YuXiang frowned.

"There may be another way. I'll write to him and send him a gift. I'll tell him I'm willing to help him if he would like."

"Good idea."

Anton wrote the letter immediately, and with the help of a friend sent a new bow as a gift. He also sent many photos of himself, his family, and, of course, his violins.

A few months later, towards the end of 1999, two young Indonesian students knocked on Anton's door. They presented him with a letter, and a large package that he found contained a badly damaged violin. The students chatted for a few minutes, but were clearly keen to get on with seeing the sights of Hong Kong.

"I'll translate the letter for you," Anton said to YuXiang, as soon as the door shut behind the Indonesians. "I've wanted to know what happened to him for so long."

YuXiang folded her knitting and settled back to listen.

"Dear Anton Sie,

"On June 8th I received your letter. On June 15th I received your gift of one professional bow, brought by my student Lili.

"How happy I am for your uncommon success and achievement, especially in the area of violin making and repair.

"My story is not so smooth yet in 1964 I got a scholarship to study once again in Germany in the *Statliche Fachschule fur Geigenbau*. But due to the G.30.S [a 1984 Indonesian docudrama film suggesting the Sep 30, 1965 coup was orchestrated by Communists] all attempt fail and . . .

"1960-1975 I was still working in Radio Semarang as Director of the Radio Semarang Orchestra and I gave private violin lessons.

"I moved to Jakarta. 1976-1980 I have been the manager of a violin making factory and supervisor of music teachers.

"1981-1986 worked as violinist in Jakarta Symphony Orchestra.

"1986 and after I gave private violin lessons and made repairs. I have just read the book of Bachman and Heron Allen.

"I believe that Anton Sie is more advanced than Anton Piontek, and this makes me very proud. I thank you for your big gift of the excellent bow which will help me perform heavier music.

"I have seen your beautiful violin photos and I am really very excited and surprised.

"I will stop my letter here. If there is any change with your address, please let me know. Thank you so much.

"With best regards

"Anton Piontek."

Anton sat lost in thought for several minutes. YuXiang quietly watched him.

"He was a good teacher, an excellent violinist. Look, he says he 'once again' hoped to study in Germany. He must have been a German Jew, like people said. But somehow he did not have the good opportunities I have had." Anton sighed.

"We have been fortunate," said YuXiang, after a pause.

"It's strange. I can't quite imagine Mr. Piontek being involved with a violin factory. Perhaps he really needed money. A factory can make student violins, but not masterpieces. Someone once asked me if we could could use science to make good violins by machine. But great violins belong to art. It is good to use science to discover the secrets of the master luthiers, but art is beyond science. A *Mona Lisa* made by a machine would just be an image. Robots can be useful, but they should be tools. To take away human art, to let machines take over, would be such a sad, even wicked thing. Printing has advanced greatly, but how can it remove art? Ah, we must know what are the limits of science and machines, and not go too far!"

After some silence, YuXiang looked up. "Like that bomb they dropped on the Japanese cities and destroyed everything?" she suggested.

Anton raised his head, startled, and nodded agreement. "Yes, like that."

After more minutes lost in thought, Anton finally took a deep breath, and went over to the package the students had brought. He gave a gasp when he saw the state of the violin inside. "I don't think any machine could fix this violin!" he exclaimed. "Perhaps that's why I enjoy violin repair. Sometimes it is even more demanding than making a new instrument! It takes all my art, science, and ingenuity to fix some problems."

For weeks Anton worked on repairing that old violin. There were times when he almost despaired of ever getting it restored. The neck especially was badly damaged, and required extremely careful restoration. But at last it was back in good working condition, and Anton felt the instrument had come alive. He packaged it most carefully, and sent it to Jakarta with a trustworthy business friend. Inside the package was a note saying Anton had done the restoration free of charge as an expression of his appreciation for all that his teacher had done for him.

Anton received a friendly letter notifying him that Mr. Piontek had received the violin. But a disappointed Anton heard no more from his teacher.

Then one day an Indonesian friend returned from a visit to his homeland.

"You knew the violinist Anton Piontek, I believe," he casually said.

Anton was all attention.

"It's sad, but I read in the papers that he died of cancer sometime in 2000. They say he was great musician, and was making music almost till the end of his life. Seems he was an interesting person. He was clearly European, but he behaved just like an Indonesian, and there was much speculation about where he originated. Some even say he was a Jew."

A wave a great sadness flooded Anton, as he realized his teacher could have used the lovingly restored violin for only a few months. But then he recognized something far more important: the violin had been repaired, his teacher had received it and thus certainly knew that he was greatly appreciated. So all the work of repairing had been worth it after all. This work was truly art, the art of appreciation and love, and no machine could ever do what he had done for that old violin of his revered teacher.

*****

In 1989 Anton was thrilled to be invited to attend the International meeting of the Cat Gut Acoustical Society, ISMA, held every four years. It gave him opportunity to meet and learn from world authorities in violin construction, but especially the chance to share his own findings on plate tuning. The conference was held that year in Mittenwald, Germany, home of the violin-making school *Staatliche Berufsfachschule für Musikinstrumentenbau Mittenwald,* founded by Matthias Klotz about 1685. The school's location (the town's name literally means "in the middle of the forest") ensured a good supply of top quality tonewood. Anton watched as trees in the forest were carefully selected for violin making. He was interested in the more mechanized methods of construction used at the school, but was not tempted to abandon his own techniques.

The lectures were technical and complicated, sometimes hard to understand. Anton suddenly understood why Europeans believed the art of good violin making had been lost! The scientific data was interesting, but Anton suspected some speakers were more intent on maintaining the mystique of violins, than unfolding their secrets.

The presentation of his own findings regarding Bitri plate tuning was the highlight of his life, because after the lecture Carleen Hutchins turned to the audience and spoke of her support for his work and methods. This was the first time since their meeting in China seven years previously that she had publically endorsed his findings, despite the considerable correspondence between them. Her complete endorsement did not come till several years later, but it was at this conference that Anton first realized this world renowned authority appreciated his conclusions and construction methods. When in 1998 Carleen Hutchins fully endorsed his findings and methods, his life-long dream came true. He had helped discover a key secret of a Stradivarius violin.

In 2001 Anton was asked to lecture at the Xinghai Music Conservatory, in Guangzhou. This gave him opportunity to talk about the history of violin making, and his own research into plate tuning, as well as the research of the Cat Gut Society in the USA. This time he could state his work was endorsed by Carleen Hutchins.

In the summer of 2004 Anton was invited to speak at the Shanghai Violin festival, that year held two hundred kilometers south in the Yandonshan, "Wild Goose Pond Mountain," resort. This area is of great natural beauty and cultural value. It has many vertical rock faces, pinnacles, mountain slopes covered with lush forests, bamboo forests, and some spectacular streams and waterfalls. It also has ancient Buddhist temples which, despite some destruction during the Chinese Cultural Revolution, were being restored. Anton was delighted to visit such a beautiful place and expound on his knowledge of violin construction.

Both these speaking appointments gave Anton the chance to share his belief that beautiful violins do not need to cost a fortune. Old Italian violins, whilst wonderful works of art and handsome pieces of preserved antiquity, are not the only ones that have stunning acoustical qualities. Numerous blind concert tests have shown that it is impossible by sound alone to differentiate old from top quality new instruments; often the winner regarding tone has been a new violin.

These invitations, despite Anton's modesty, indicate his assured standing as a violin expert in China and Hong Kong, where his experience and opinion is highly regarded. Perhaps, he hopes, even in Europe the science of violin acoustics will be recognized. Meanwhile, he is content to know he has done his best to share the truth about these beautiful instruments. He smiles: what began as a background hobby for his primary training as a physicist has become the main focus in his life, and his primary contribution to posterity. He hopes that not only will his excellent violins bring joy to many people, but the techniques he discovered will allow others to make beautiful violins.

Mr. Carl Pini's visit for repairing his Carlo Bergonzi (1732), (1982 Hongkong.)

Mr. Pini played Anton Sie Violin #50.

Carl Pini and his 1732 Carl Bergonzi violin that Anton repaired and identified.

The Don Pasquale opera, when Anton played solo guitar with tenor Brown Bradley. The lower R photo shows Anton, L, with the other guitarist, Lo Chi Keung, and Brown Bradley.

During a break on the lecture tour of China, 1982. L-R Sie Anton, Prof Chen Qiang, Morton Hutchins, Carleen Hutchins, and Prof Wang Xiang.

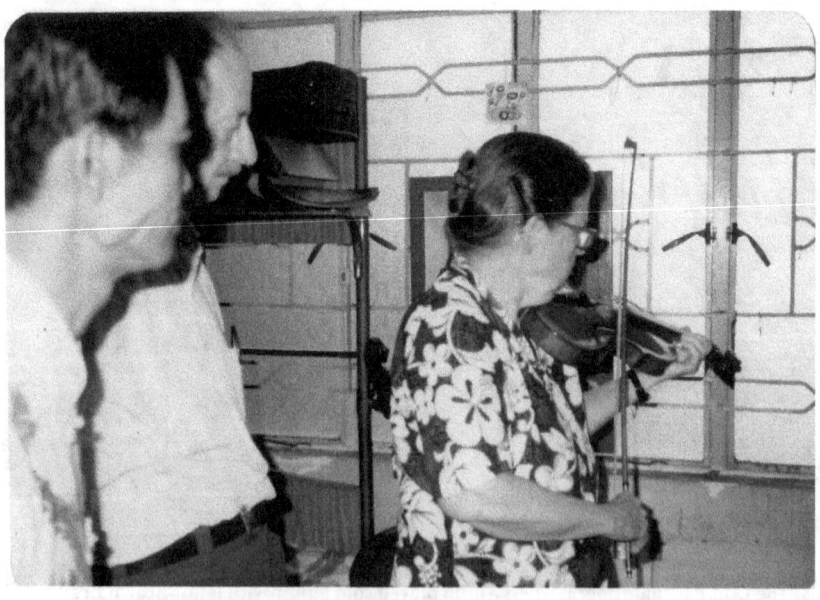

Carleen Hutchins testing one of Anton's violins, while he watches anxiously, as his tense jaw muscles show.

Students' Party, 1986. In upper L photo NeeZi, far L, plays beside Luiji Oliva and Sven Östring. In upper R three young students play. In lower L photo Lui and Sven play together.

ARTE LIUTARIA
DICEMBRE 1988 - N. 12

Gentile collega
Abbiamo letto con molto piacere la sua lettera, la ringraziamo anche per la cifra che ci ha spedito. Abbiamo visto dalle foto riproducenti i suoi strumenti che lei ha raggiunto un ottimo livello e riesce molto bene ad esprimere nella liuteria la cultura e il gusto delle arti del suo paese.

Dear colleague,
I read with great pleasure your letter and thank you for the contribution you sent us. We have seen from the photographs you sent that you have reached an excellent level of work and succeed in expressing through violin making the culture and taste of your country's art.

著名的提琴制造大师，《意大利提琴制造的经典线条》作者及"提琴制造艺术"杂志编辑卡罗-维多利在该杂志的公开信1988年 第12期译文

敬爱的施安顿同事：
我很有兴趣的读了您的来信，谢谢您的支持。我从您寄来的相片中看到您的技术是达到了卓越的水平而且能把您国家的文化艺术和品味通过小提琴的制造很成功地表现出来。。。。。

克雷蒙娜和中国古老藝術完美的结合．

Presented by Hongkong Art Center 25 Oct- 17 Nov 1990.
香港藝術中心主辦．一九九零年 十月二十五至十一月十七日．

The Dragon Violin Exhibit.

CATGUT ACOUSTICAL SOCIETY, INC.
112 ESSEX AVENUE   MONTCLAIR   NEW JERSEY 07042   USA
FAX: (973) 744-9197   EMAIL: CATGUTAS@MSN.COM

September 3, 1998

President
  J. Maurits Hudig
Executive Vice President
  Julius VandeKopple
Treasurer
  A. Duncan Kidd

Permanent Secretary
  Carleen M. Hutchins
Executive Secretary
  Elizabeth McGilvray

Research
  Oliver Rodgers

CAS Journal
  A. Thomas King, Editor
Associate Editors
  Gregg T. Alf
  Robert T. Schumacher

Musical Acoustics Research Library
  Joan E. Miller

Violin Octet
  Carleen Hutchins
  Marina Markot

Vice Presidents of the International Committee
  Herman Medwin, Chair
Australia
  John Godschall Johnson
Canada
  Warren Reid
France
  Voichita Bucur
Hong Kong
  Anton Sie
Italy
  Domenico Stanzial
Japan
  Isao Nakamura
The Netherlands
  Adrianus J. M. Houtsma
Sweden
  Anders Askenfelt
United Kingdom
  James Woodhouse

Advisory Council
  Donald Engle
  Dennis Flanagan
  Frank Lewin
  Yehudi Lord Menuhin
  Harry S. Wake
  Gabriel Weinreich

Mr. Anton Sie
No 1 Babington Path, 8M2
Hong Kong

Dear Anton Sie,

I am very much interested in your Tri Harmonic Matching of violin plates and the result you are getting. Years ago, when I tried this, I found the results were hard and shrill for the viola so have not recommended it. I suspect the modes were not too well tuned to start out with. I am delighted you find that the Tri Harmonic Matching works so well. We are sending your paper to the editor of the Journal for review. Thank you for this.

Morton is still recovering very slowly from his stroke. I am fine and am working on a violin (#478) as well as five violas. We have had a good summer in New Hampshire and send you our best wishes.

Sincerely,

Carleen M. Hutchins

CMH:MG

Carleen Hutchins' letter that finally gave Anton recognition for his contribution to the understanding of violin plate tuning. Note Anton is designated as a vice president of the society.

施氏五弦提琴
Sie's Model Pentaline

　　在樂器制造史上曾有一些制造大師制造過一些五弦提琴,但始終没成功。主要是因爲音色和演奏上的問題。他們多數是從中提琴或小提琴出發進行設計。制成的五弦琴音量小或者很難拉奏(參看 C. 波尼亞托夫斯基《中提琴藝术史　》)。

　　本人采用了諧和對等方法,以甘巴琴爲基礎制造了嶄新的五弦提琴,引起許多中小提琴界的關注,其特點是具備了中提琴和小提琴的優點而且琴體不大,易于拉奏,可以當小提琴或中提琴兩用琴,因而夸大了普通四根弦提琴的音域。

**　С. Понятовский　　История альтовото искусства

Anton Sie Pentaline 2000#134　　五弦琴《　鍾馗　》

The *Pentaline*, Anton's five-stringed violin.

Photo of YuXiang and Anton, 1986. YuXiang looks radiantly healthy, and triumphant: at last she got Anton to a party! After years of hardship, even Anton could finally relax and enjoy life occasionally, although he looks a little self conscious in a formal suit!

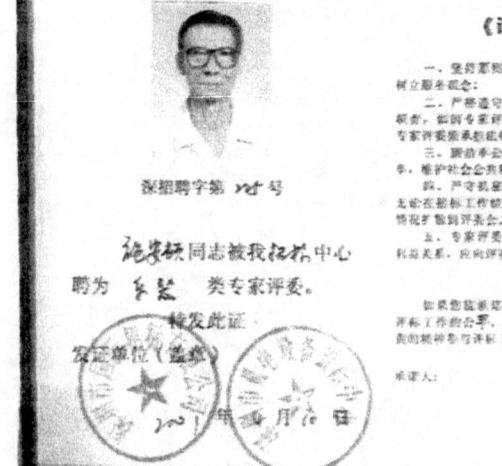

Man of the year.

Postlude

The Guitar, in Appreciation

THE BEAUTIFUL SOUNDS OF Jenő Hubay's *Der Geigenmeister von Cremona* died away, and Anton reverently placed the violin on his workbench. He had just completed his third tri-*Eigen* tone matching violin. His favorite test-piece sounded wonderful.

He stood up, and came out to YuXiang.

"I've done it! I've done it again!" he announced triumphantly.

"Great!"

"I feel so good, but it's very demanding. I never realized how much energy it drains from me. I can consistently make *bi-Eigen* tone violins, but getting that third tone to match is extremely hard."

He went over to the coffee table, picked up his old guitar, and idly plucked the tune of *Der Geigenmeister* along its strings.

YuXiang looked up from her knitting. "It's strange, but violins have fascinated you all your life. You've devoted so much time and thought and energy to them. But when you want to relax and have fun, you always turn to your guitar."

Anton stopped playing, and a bewildered expression clouded his face. Lost in thought, he gazed at his wife for several minutes.

"You're right!" he finally announced. "The violin is my master. It drives me to perfection. I've been fascinated to learn its secrets. But the guitar is my friend."

YuXiang smiled. A gentle silence settled between them.

"You know, he made it so easy. I mean, Mas Toha. Music just flowed out of him. He could make anything sing. Even my cranky aunt's cooking pots!" Anton smiled at the memory.

The rhythmical clicking of YuXiang's knitting needles broke the stillness.

"Perhaps I should let myself relax a bit more. Mas Toha told me he played *gamelan* because he loved it. He earned some money but he played because he couldn't help himself. Anyway, he made a good living from his pot mending, so the *gamelan* really was just for fun."

YuXiang placidly continued knitting.

"Like you!" announced Anton, with a sudden realization. "You knit just because you like it, don't you?"

"True," agreed YuXiang. "Perhaps it was a chore once, but now I make these bears just for fun."

"I thought you did it for that children's charity."

"Well, you have to have an excuse!" laughed YuXiang, her eyes twinkling.

"I guess so. Remember that wonderful excuse I had for playing guitar, the one Carl Pini gave me? When I played the Don Pascuale opera? I'll never forget that! I finally felt a respected person in Hong Kong. And think what I earned!" Anton paused, savoring the memory. "I'd never had so much money in my life! And yet it all began with just having a bit of fun."

YuXiang's needles clicked away.

"Yes, and remember Jan van den Berg? How much I owe him! But did I tell you, I discovered recently that when he went to New Zealand he became the teacher of Sven Östring! Fancy that! We live in such a small world." Anton shook his head in wonder.

Suddenly the strident call of the telephone broke the stillness.

Reluctantly, Anton got up and answered it.

"So you've got another one to make?" YuXiang said when he returned.

"Yes, I have another order for a violin. Strange, when I don't need more orders for money, I get more than I can cope with. But I will always enjoy the challenge."

YuXiang counted her stitches. She was making the face of a bear, and needed to be very accurate. She looked up. "And you never can say no to helping someone, can you?"

"I suppose so," Anton nodded sheepishly. "But as I was saying before the phone rang, I'm going to allow myself more relaxation."

"Really?" his wife responded skeptically.

"Actually, I've been playing with the guitar for years," Anton admitted hesitantly. "I've been writing music for it, *gamelan* style music."

YuXiang raised her eyebrows and opened her eyes wide. "Really?"

Anton nodded. "I've worked so hard for years. Sometimes I just had to take a break. I miss Indonesian music."

"If course."

Anton went to his workroom and brought out an untidy sheaf of papers. "It's all a bit disorganized," he said apologetically. "But it wouldn't take much to turn it into something worthwhile." He laid his manuscripts across the coffee table.

"Remember the fun we had going to NeeZi's graduation!" YuXiang exclaimed.

Anton looked up, surprised at this irrelevant remark. "You enjoyed it?"

"Yes, I did, but as you know I found travel wasn't good for my health. The doctors at Ruttonjee did a good job, but I don't have perfect lungs. But what I was thinking was, why don't you visit NeeZi again? She's always asking us, and it would be a nice break for you."

Anton nodded and looked pensive. "Yes, I could do that. Remember the trips I made in 1989 and 1990? Ah, that ISMA conference in 1989, that was amazing, the highlight of my life! And the Lyons festival of violin making was interesting, although I'm not sure what Stradivarius would think of their methods."

"But those trips were for work. This would be for relaxation."

"NeeZi's graduation was supposed to be relaxing, but you got exhausted!" retorted Anton.

YuXiang nodded. "You were all so determined I should see all the famous places, but we couldn't see everything. It was so frustrating. I just wanted to see where NeeZi lived and worked, but you made me see all those violin places!"

Anton frowned and was silent.

"I could see NeeZi and her family, and go to Spain and look at guitars. It would be a different sort of celebration this time." He sat deep in thought.

YuXiang smiled. "If you're talking about celebrating, why don't we go out for a meal tonight?"

"Why not!" laughed Anton. "We'll celebrate my tri-matching violin together here, and then I'll go to Europe and celebrate there."

So Anton went to Europe. He visited NeeZi in Vienna, and went to Cremona (again) and on to Madrid. He enjoyed Spain with its unique history. Europeans had expelled their Moorish conquerors, but evidence of their presence was seen everywhere in Spanish architecture, reminding him of his Indonesian childhood. Music shops drew him like a magnet. The highlight of his trip was finding the renowned workshop of Angel Benito Aquado in Madrid, where he bought a beautiful *Sor* (Romantic) guitar which he used for his logo ever since.

Anton's renewed interest in guitar rekindled his interest in Chinese instruments. He played a *pipa* in China, demonstrating its similarity with the guitar. Now, as he began composing Chinese music for guitar, he decided, at the age of seventy-three to learn the *guzheng* (Chinese zither). With a history going back at least two and a half thousand years, the *guzheng* is still one of the most commonly played Chinese instruments. Anton took regular lessons from Miss Zhang LinLi, and two years later in 2010 he passed grade six level of *guzheng* performance with the Beijing Central Music Conservatory. His purpose was primarily to apply *guzheng* tonal and technical qualities to the guitar, enabling him to write and play classic Chinese music on his guitar.

In 2009, the year he was deeply saddened to learn of the death of Carleen Hutchins at the venerable age of ninety-eight, he completed a collection of guitar compositions: *Pluck and Play Traditional Chinese Music on the Guitar*, followed by *Indonesian Music for Classic Guitar* (2013) and *Arabian Music for Classic Guitar* (2013). Anton was delighted to discover Facebook in 2004, and realizing its potential, he placed his guitar compositions on it, so anyone can freely download and enjoy his music.

"So many people from so many places have helped my music. Now I want to share as much music as possible with others," he says.

*****

Anton was happily loading classic music from YouTube on to his website one morning in July, 2016, when the phone rang. From being slightly annoyed at the interruption he was all attention when he heard the voice at the other end.

"This is Ong BingGian's son," he heard. "I thought you should know that my mother passed away last week."

There was a lot of static on the phone, and Anton was wrapped in silent thought.

"Hello! Hello! Hello! Can you hear me? Are you still there?" The phone crackled.

"Yes, I am here. But I am shocked."

The phone continued to crackle, and Anton had difficulty understanding the voice from faraway America.

"Can you hear me?" asked Anton, as the static continued to drown the voice of the caller. "I'll give you my email address, and you can contact me that way."

Anton put down the phone, and sat lost in memory and grief.

When YuXiang entered the room she knew instantly that her husband had received bad news. Quietly she made a cup of tea, placed it in front of him, and left the room. For two days Anton was wrapped in sorrow.

On the third day, as YuXiang was quietly placing a plate of noodles in front of her husband, Anton's fog of misery suddenly cleared. Unexpectedly he smiled.

"What an idiot I am!" he exclaimed. "Such an idiot! Why am I grieving a schoolgirl I loved in Indonesia more than sixty years ago!"

He was silent. Chinese are not demonstrative. But he had to say it, just this once. "YuXiang, thank you for standing behind me through everything. I will always love you!"

"I know," she said softly. "I know you do. And it is good to hear you say so!"

"I wish, I really wish," and Anton took a deep breath. "I wish I had tried to get you a ride on the bus, instead of letting you walk in the snow."

"Really!" exclaimed YuXiang, sitting down suddenly. She opened her mouth, but no sound came out.

Anton's Luthier Logo, with his Romantic *Sor* Guitar from Madrid.

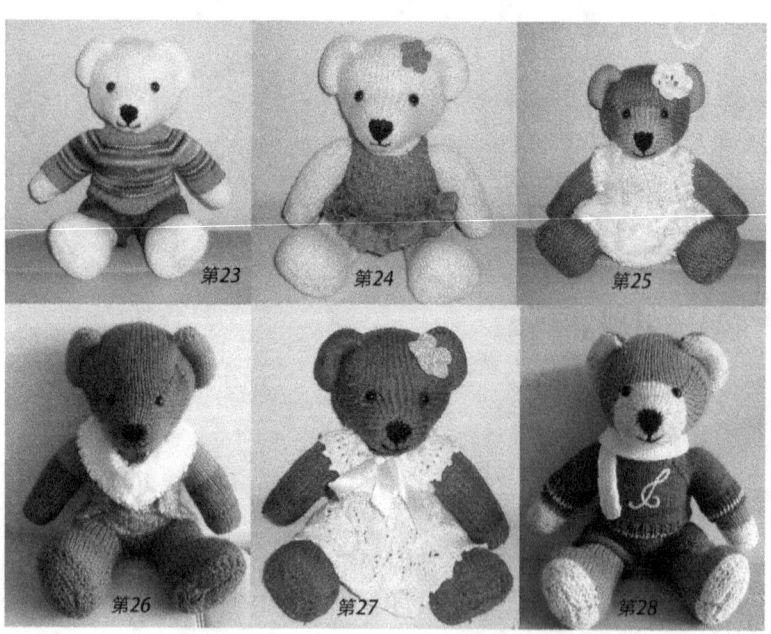

Some of the bears YuXiang knits for charity.

YuXiang and Anton celebrating their Golden Wedding anniversary on the balcony of their home, 2012.

YuXiang and Anton ready for Christmas, 2013.

One of Anton's many guitar compositions.

Appendix A

How to Make a Good Violin

ANTON SIE'S DESCRIPTION (WRITTEN especially for this book) of his method for making a good-sounding violin.

How to make good sounding Violin

In the violin body, the top plate and the back plate are the two most important vibration bodies. each of them has their own that so called Eigen frequencies (or characteristic frequencies). According the theory of resonance, two bodies with the same frequencies will resonant and give the most big vibration of the system.

1. Tune the Eigenfrequencies of the Top plate of the violin such that Equal to their correspondent Back plate's Eigen frequencies.Get thew loudest sound.
2. Tune the eigenfrequencies such that the are in Harmonic Series. Produce Rich Sound.
3. The most important modes are the 5th,2nd and 1st mode of the Eigen Frequencies. Often called as O ,ode, X mode and X mode respectively.

How to make good sounding violins

Prepare well crafted violin plates (Top and Back) the outer surface either varnished or keep white. the thickness distribution approximately like this

Fig. 1

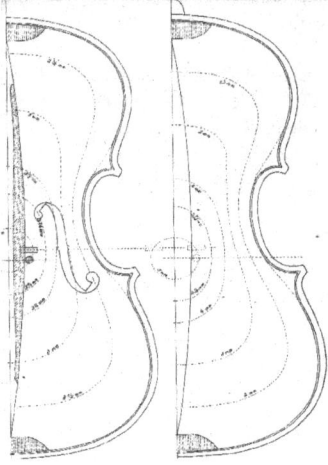

Start with the back plate
 Handle with fingers on point h or h1h2 , tap with finger on the middle of the plate such that can clearly hear the tap tone

The handling of taping mode 5

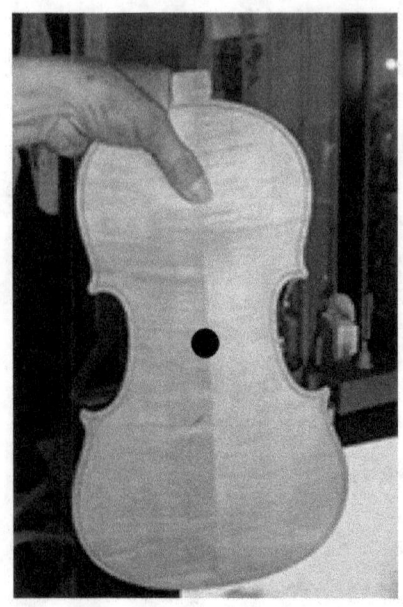

H1 Handling

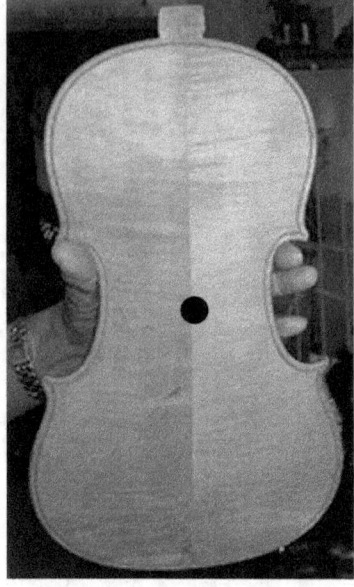

H12 handling

Tap with fingertip on the black circle area, we can hear the tap tone. This is the tap tone of mode 5, the strongest one !(Mode 5 Eigenfrequency)
By thining the plate's thickness we can lowering the tap tone.

Similar procedure with the Top plate (with bassbar installed) such that the Eigen frequency of the both plate have equal value. This called matching frequency of the mode 5. For example matched with mode 5 value of 340 hzs for each other. After we assembled the violin, we got that so called one to one matching violin with mode 5 frequency of340 hzs. We will get pretty good sounding violin.

It should be noted that listening tap tone is not so easy !!!!!!! But using the electronic measurement is easy and accurate yet !

Need : powerful amplifier , audio frequency generator and mid frequency high capacity speaker. 4 sponge pieces , powder tea (tea bag !) or granule wood powder.

Put the 4 sponge pieces such that approximately touch the underside of the 4 wings (bee stings) of the back plate of the violin, spread the tea powder on the upper surface of the plate, scan the frequency generator slowly, at the range of 340 Hzs, See fig.

Find the frequency that the tea particle jumps highly. Stop scanning. Wait. till the tea particle stop to move. This frequency just the Eigenfrequency of mode 5. By thinning the plate uniformly, we can lower the frequency. The suitable frequency is between 320 to 360 hzs.

Do the same procedure with the Top plate (with the bass bar fitted). By thinning the inner surface and the bassbar the frequency will drops. Tune to the same frequency with that of the back plate. This called matched the 5th mode to equal Eigen frequency.

Assembling the well tuned plates. After all fitting were done. We will got A well sounding violin,

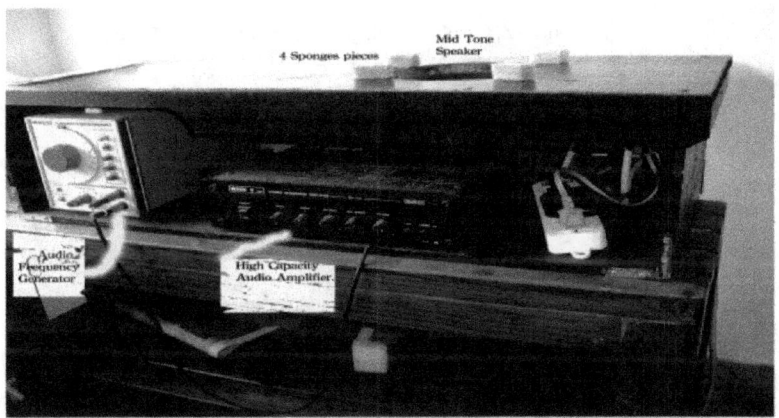

Fig 2

Using this devices we can detect and tune more Eigen frequencies of mode 2 (X mode) and mode 1 (+ mode) using the arrahgement of Fig 3. N are the places of the supporting sponge pieces. And A are the places of the speaker.

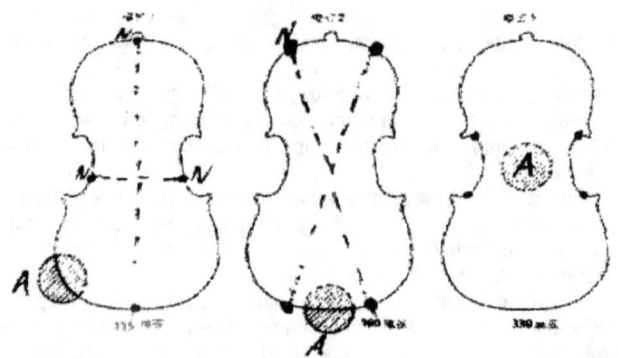

Arrangement of measurements: Right for mode 5, middle for mode 3 and left for mode1

There are many eigen frequencies of the plate, the Eigen frequency of mode 5, mod2 and mode 1 are the most important frequencies.

Nomenclature

If the one corresponding Eigenfrequencies of the Back and top plate are equal is called mono matching.

If two coressponding Eigenfrequency of the Back and Top plates are equal respectively it is called Bi matching.

If three of the corresponding Eigenfrequency of the Bavk and Top plate are equal respectively it is called Tri matching.

It should be noted that it is possible to tune the frequency of mode 5, mode 2 and mode 1 of a plate such that in octave relation or in other words ,in a harmonic serries. Very difficult but possible such kind of arragement of frequency matching is called "Harmonic Matching"

Example of Trharmonic Matching Plates.

Eigen Frequencies of Top Plate : 90 Hzs (mode 1 or mode +). 180 Bzs (mode 2 or mode X) 360 Hzs (mode 5 or mode O).

Eigen Frequencies of the Back plate: 90 Hzs (mode 1 or mode +). 180 Bzs (mode 2 or mode X) 360 Hzs (mode 5 or mode O).

This is that so called Triharmonic Matching with O=350 Hzs.

Appendix B

Carleen Hutchins's Correspondence

A FEW SAMPLE COPIES of the voluminous correspondence between Anton and Carleen Hutchins regarding his Bi-Tri harmonic plate tuning technique.

150

CATGUT ACOUSTICAL SOCIETY, INC.
112 ESSEX AVENUE   MONTCLAIR, NEW JERSEY 07042

August 1, 1991

Mr. Sie Anton
No. 1 Babington Path
Eighth Floor M2
Midlevel, Hong Kong

Dear Mr. Sie:

Many thanks for your recent letter. I am delighted to have your further information on the Bi-Tri tuning. Your findings that mode #2 frequency stays higher with the long flat arch fits exactly with our findings. Perhaps you should write a short article for the Journal on this.

It is good to know you are continuing to make fine instruments. Do let me know when you get some results on the A1 B1 Delta.

Keep up the good work. Morton joins me in sending you our very best wishes.

Sincerely,

Carleen M. Hutchins

CMH:EM

President
 Joan E. Miller
Treasurer
 Paul B. Ostergaard
Permanent Secretary
 Carleen M. Hutchins

Vice Presidents
 Oliver Rodgers, Finance
 Vochta Bucur, Europe
 Norman E. Hannan, U.K.
 Erik V. Jansson, Scandinavia
 John G. Johnson, Australia
 Richard Sacksteder, U.S.A

Journal Staff
 Daniel W. Haines, Editor
 Sydney Fox
 A. Thomas King
 Edward Wolf
Executive Secretary
 Elizabeth McGivney

151

CATGUT ACOUSTICAL SOCIETY, INC.

112 ESSEX AVENUE   MONTCLAIR, NEW JERSEY 07042

September 20, 1991

Mr. Sie Anton
No. 1 Babington Path
Eighth Floor M2
Midlevel, HONG KONG

Dear Mr. Sie:

Many thanks for your letter early this summer. I am glad to know you are busy making and repairing violins.

Your continuing experiments with the Bi-Tri tuning methods are very interesting, especially as it relates to different models, different arching contours and different graduations. I am finding the same things that you are relative to the arching shapes - particularly that the long flat center-type arching of the top plate makes it easier to keep mode #2 frequency up. I have not made a violin with the Stainer arching as yet, but some of our tests have shown that the arch height in the upper and lower bout areas is critical for cross-grain stiffness and keeping mode #2 frequency high.

It would be interesting to have you write something of your findings along these lines when you feel ready. Also your findings on the A1-B1 Delta are along the same lines as my work, but there is a lot to learn in this area.

Morton and I are having a good summer here in New Hampshire and we both send you and your lovely wife our very best wishes.

Sincerely,

Carleen M. Hutcyins

CMH:EM

152

Antonio Sie
Science Arts Liutaria
MUSICAL INSTRUMENT ENGINEER
J.M.I.E., Bs. T. Phys.)
Member of the CATGUT Acoustical Society, Inc. New Jersey
and the Acoustical Society of America, New York
EXPERT IN VIOLIN MAKING &
STRING INSTRUMENT REPAIRING

No. 1 BADINGTON PATH 8TH FL. N2.
MIDLEVEL HONG KONG
TEL: 5492292

p.1

26 Nov 1991

Dear Mrs C.M. Hutchins

Glad to receive your nice letter, I am happy if Mr Rodgers will be interested with my result and you can show my last letter to him, if he would like to know more about it, I am happy to do with my best for his plan.

Concerning the unorthodox (or irregular) graduation, I have paid attention a long long time ago, I have got some hints from the repair of old violins, especially the cheap rude Mittenwald fiddles... Later I examined also the Italian masters violins and some literature's data. I do some improvements by adding horizontal brace for the old violins those I have very low unbalanced X mode.

① Along the grain direction of braces will raise O mode more than the other

② Horizontal ~~parallel~~ braces will raise X mode more

③ Slinting braces will raise both X & O depending on the angle or

Till 1988~1989 I applied all of these things to my Dragon Violin (the irregular crafted Dragon acts as the braces — self cut brace —) for its back and able to get Biharmonic on the back plate. I applied also to the repairing work for the cheap old violins which are too thick (plenty of wood). Due to the limit

153

Antonio Sie
Sciene Arts Liutaria
MUSICAL INSTRUMENT ENGINEER
·M.I.E., & T. Phys.I
Member of the CAS ?T Acoustical Society for New Jersey
and the Acoustical Society of America Inc. York
EXPERT IN VIOLIN MAKING &
STRING INSTRUMENT REPAIRING

8a 1 BABINGTON PATH 8TH FL M2
MIDLEVEL, HONG KONG
TEL 5490282

p 2

of outer finished shape of old violins some time it is very hard to tune, even in Biharmonic or one to one matching. If the X mode to high (or hard to lowering) I scrapt in the perpendicular (along the grain) direction unevenly (fig ①) left 4~6 strips on the upper and lower bout area, the thickness of the strip c.a 0.3~0.5 mm width c.a 8mm and smoothing to even graduate near the border and middle bout. Such kind of irregular graduation will help (even not always 100%) to lower the X mode garden.

Cross section (enlarged)

If the O mode hard to drop I scrapt like ② in horizontal direction.
Tonal quality quite satisfactory. The horizontal strips improved the low X violin very well. The G and D string more powerful and eliminated the "hollow" low strings sound of low X violin. It seems reinforce the horizontal tension to resist the rocking motion of low string bridge/bassbar vibration.

154

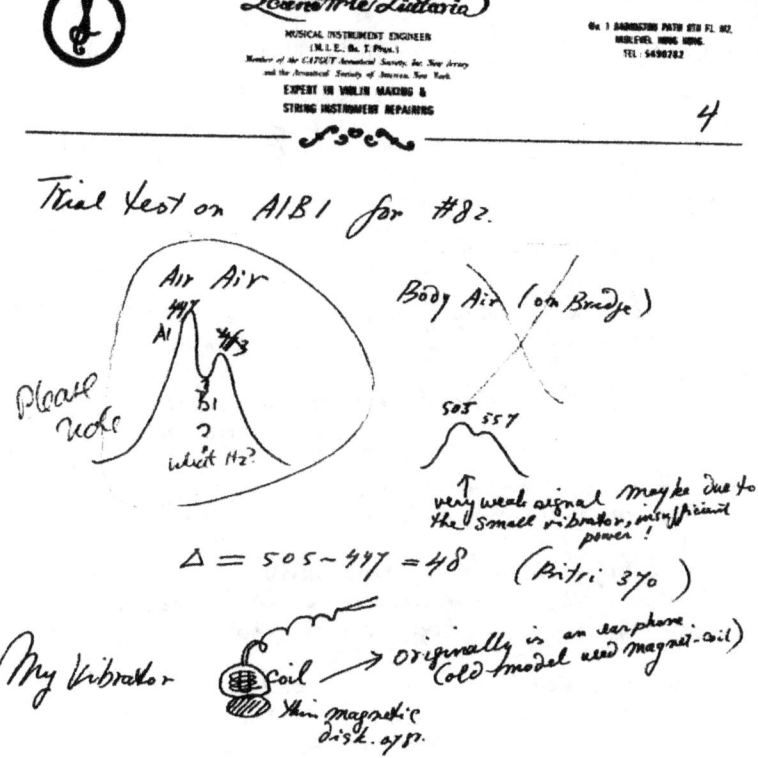

Appendix C

Teacher Anton Piontek

LETTER (IN INDONESIAN) TO Anton Sie from his teacher Anton Piontek.

Jakarta, 27 Sept '99

Dear Anton Sie,

Surat anda telah saya terima beserta foto² dengan senang hati pada tgl. 8 Juni 1999.

Tgl. 15 Juni '99, saya menerima satu bon dari Anda yang dibawa oleh Lili (murid saya).

Betapa girang hati saya, mendengar Anton dengan keluarga selamat dan bahagia.

Terlebih lagi melihat kemajuan Anton <u>luar biasa</u> dalam bidang musik umumnya, khususnya dalam bidang biola (membuat dan mereparasi).

Riwayat saya. Semestinya dalam th. 1964 saya mendapat beasiswa ke Jerman kota Mittenwald untuk mendapat pendidikan dalam pembuatan dan perbaikan biola di Staatliche Fachschüle für Geigenbau.

Tapi karena ke.dahuluan adanya G.30.S., surat² saya di Dep. Luar Negri menjadi hancur dan berantakan, jadi rencana saya gagal karena menteri dan pegawai²nya diganti. Saya menjadi putus.asa, dan tidak meneruskan

1.

Di th. 1960-1975 (15 th.) saya hanya menjadi dirigent Orkest Radio Semarang dan memberi privat biola.

Pada th. 1976-1980 (4th.) saya menjadi manager pabrik pembuatan biola di Pontianak (kalimantan) dan supervisor guru² yang mengajar musik.

Pada th. 1981-1986 (5th.) saya menjadi pemain biola Orkest Symphony Jakarta. RRI.-Jakarta. Orkest tersebut bubar, karena dirigentnya korupsi.

Dewasa ini saya mengajar di sekolah musik, memberi privat dan mereparasi biola di rumah.

Modal saya untuk mereparasi hanya buku Alberto Bachmann dan Heron Allen. Sedangkan alatnya saya terima dari seorang Pastor tapi tidak lengkap. Tapi saya pakai lèm dari Eropa.

Dapat saya pastikan bahwa Anton Sie jauh lebih maju dari pada Anton Piontek. Dan inilah kebanggaan saya.

Dear Anton Sie. Maafkanlah saya, karena surat balasan saya datang lama sekali disebabkan keadaan.

2.

Terima kasih saya yang tak terhingga untuk bow yang saya terima dari Anda.
Dengan bow tersebut saya dapat main lagu: yang agak berat.
Foto: yang saya terima menunjukkan bahwa anda sudah dapat membuat biola yang amat bagus. Saya sungguh mengagumi anda.

Apabila anda pindah alamat, sudilah memberitahu pada saya.

Nah sekian dahulu, sambil saya menunggu nasehat yang mungkin anda dapat berikan kepada saya.

Salam hangat saya kepada anda sekeluarga

Dari,

(Anton Piontek)

3.

www.ingramcontent.com/pod-product-compliance
Lightning Source LLC
Chambersburg PA
CBHW070248230426
43664CB00014B/2452